Ethical Life

ETHICAL LIFE

ITS NATURAL AND SOCIAL HISTORIES

WEBB KEANE

PRINCETON UNIVERSITY PRESS

PRINCETON AND OXFORD

Published by Princeton University Press, 41 William Street, Princeton, New Jersey 08540

In the United Kingdom: Princeton University Press, 6 Oxford Street, Woodstock, Oxfordshire OX20 1TR

press.princeton.edu

Cover art by Webb Keane, photographed by Clara Keane. Courtesy of the author.

Cover design by Leslie Flis

Second printing, and first paperback printing, 2017
Paperback ISBN 978-0-691-17626-0
Cloth ISBN 978-0-691-16773-2

Library of Congress Control Number 2015942446

British Library Cataloging-in-Publication Data is available

This book has been composed in Sabon Next LT Pro with Trajan Pro Display.

Printed on acid-free paper. ∞

Printed in the United States of America

For my teachers and students

Contents

ACKNOWLEDGMENTS

Like ethical life itself, even the solitary business of writing is thoroughly enmeshed with the voices of other people. Here are just a few of those who lurk in these pages. Matthew Engelke, Don Herzog, James Laidlaw, and Adela Pinch read the entire manuscript at least once, each with unsparing acuity. In Adela's case, this has been yet another conversational turn in a dialogue that for decades has influenced my thinking about, well, pretty much everything. Jonathan DeVore, Paul C. Johnson, Benjamin Lee, Michael Lempert, James Meador, Elizabeth Povinelli, Greg Urban, and Michael Warner commented on significant parts. Ranging so far afield from one's specialty poses special risks; I received timely critical readings of relevant passages from Elizabeth Anderson, Didier Fassin, Susan Gelman, Lynn Hunt, Judith Irvine, Peter Railton, Joel Robbins, and Victor Lieberman; the usual disclaimers about what I did with their advice apply even more than usual. Penelope Eckert, Robin Edelstein, Nick Enfield, Sally McConnell-Ginet, David Porter, Gayle Rubin, Jack Sidnell, and Abigail Stewart responded quickly to queries in their fields of expertise. Many others have left their mark here, including Amanda Anderson, Richard Bauman, Maurice Bloch, Craig Calhoun, Victor Caston, Veena Das, Terrence Deacon, Krisztina Fehérváry, Alison Gopnik, Lawrence Hirschfeld, Charles Hirschkind, Matthew Hull, Alaina Lemon, Saba Mahmood, Alison McKeen, Mary Murrell, Fred Myers, Elinor Ochs, Alan Rumsey, Bambi Schieffelin, Scott Shapiro, Robert Sharf, Andrew Shryock, Michael Silverstein, Patricia Spyer, George Steinmetz, Kathryn Tanner, and audiences and seminar participants at Århus, Arizona, Australian National University, Berkeley, Brown, Cambridge, Chicago, Cornell, Edinburgh, Gadjah Mada, Johns Hopkins, London School of Economics, Michigan, New York University, Princeton, Toronto, the University of California–Los Angeles, Virginia, the Michicagoan Linguistic Anthropology Workshop, the Michigan Humanities Institute, the Oslo Summer School in Comparative Social Science Studies, the Center for Transcultural Studies and Institute for Public Knowledge, and the School of Criticism and Theory. A conference organized by Michael Lambek and Jack Sidnell provided an important catalyst, and Ian Hacking, some timely interventions. Parts of the following chapters appeared in earlier versions: chapter 3, in *Anthropological Quarterly* 81 (2008): 473–82; chapter 4, in the *Journal of Linguistic Anthropology* 21 (2011): 166–78; and chapter 6, in *Numen: The Journal of the International Association for the History of Religions* 61 (2014): 221–36. Fred Appel capably steered the volume through publication. John Mathias and Stuart Strange helped prepare the final manuscript. Financial and moral support came from the Center for Advanced Study in the Behavioral Sciences (Stanford), the Michigan Humanities Institute, and the College of Literature, Science, and Arts at the University of Michigan and my own department, an unsurpassed intellectual home.

PART ONE

Natures

Introduction: Ethical Affordances, Awareness, and Actions

Ethical impulses, judgments, and goals are features of everyday life in every known society, past and present. Does this mean that the propensity for taking an ethical stance arises from human nature itself? If it is innate, does it follow that we could be ethical without knowing it? There are many who would reject that idea. Some people hold that ethics is based on reason; others, that its sources are divine. If ethics is based on reason, must each individual be capable of working it out in his or her own inner thought or at least of learning from the wisdom of those who have? If ethics is divine, does this require adherence to the right laws, faith in the right gods, or consultation with one's conscience? Or is it, rather, the fact that ethics is something each society creates on its own, so that each of us is stamped with the impress of a particular tradition, borne within a specific community? And in that case, does that mean each ethical world is ultimately incomparable to any other since each is the contingent outcome of a singular historical pathway? Or does it turn out that ethics is a product of natural selection, favoring reproductive success? Does science then require us to accept that ethical concepts and values are ultimately epiphenomena, generated by mechanisms that themselves have nothing specifically ethical about them?

This book looks at several ways of answering these questions through empirical research. Broadly speaking, the approaches we will examine here fall within the traditions of either natural or social history and can lead to very different views of ethical life. Indeed, some scholars think that these two approaches are quite incompatible and insist that we must choose between them. I think that is a mistake: it is important that we are all talking about the same world. But the differences matter. Naturalistic research, in fields such as neuroscience, cognitive science, linguistics, developmental psychology, and biological anthropology, tends to seek out human universals. These often (but not always) involve processes that work beyond the scope of anyone's awareness. The research commonly (but, again, not always) takes the individual as the primary unit of explanation. It describes changes that usually unfold on the vast timescale of evolution. What I call social history includes not just the scholarly discipline of history proper but also cultural and linguistic anthropology, historical sociology, sociolinguistics, microsociology, and conversation analysis. These approaches tend to stress the diversity of existing ethical worlds. Although they often describe economic, political, and other forces of which people are unaware, they are prone to giving a central place to the agency of people who act with self-consciousness and purpose. The focus is typically on life within communities. The time frame of social change can be as narrow as a few decades.

Natural and social histories offer more than different points of view, since they challenge not just each other but also certain dominant strains of ethical thought in philosophy and religion. If some naturalistic explanations, such as seeking causes of behavior in neurophysiological mechanisms, can undermine our confidence that ethical choices are really choices, cultural relativism can seem to undermine the sense that ethics is objectively compelling or anything more than social conformity. This book argues against both kinds of debunking. It proposes that if we look closely at the points where natural and social histories converge, we can gain new insights into ethical life, the fact that humans are inevitably evaluative creatures. We might also gain something looking the other direction as well: this book also stems from the conviction that the more familiar ways of distinguishing between natural and social realities no longer serve us well and that ethics, with sources in both biological mechanisms and social imaginaries, is a good place to start rethinking their relations. With these purposes in view, this book works with a broad definition of *ethical life* to refer to those aspects of people's actions, as well as their sense of themselves and of other people (and sometimes entities such as gods or animals), that are oriented with reference to values and ends that are not in turn defined as the means to some further ends.

Researchers in the various disciplines that focus, respectively, on natural or social histories tend to stay housed within their separate silos. With some notable exceptions, they rarely take advantage of what they could learn from one another's research. Indeed, they often have principled criticisms of other styles of research, which can reinforce the idea that their findings contradict each other. The natural scientists may object that too much emphasis on social construction overlooks the objective foundations on which moralities are built. Some even suggest that resistance to naturalistic explanations betrays a lingering taste for the "supernatural." The social historians and ethnographers, in turn, worry that naturalistic explanations don't give enough credit to people's creative agency and self-interpretation, to the first-person point of view, or to the complexities and contradictions of history. In response, this book assumes that there is a lot to be gained by persuading people to climb out of their respective silos and look around.

To that end, this book brings together key findings from psychology, the ethnography of everyday social life, and social histories of ethical reform. It does not, however, aim to revive the old dream of a unified explanation for everything. It will not leap directly from genetics to social movements, say, or from game theory to theology. Rather, these chapters scout along borderlands where certain fields converge and overlap. For example, they trace out those points where cognitive science meets child development and blurs into the microsociology of face-to-face interaction, which in turn provides materials that can inspire ethical reformers working on the vast scale of religious or political revolution. The approach developed here is based on two premises. One is that both approaches, from natural and social history, respectively, provide crucial insights into ethics—I refuse to dismiss either out of hand. The second, which follows from the first, is that neither of them can provide a satisfactory account of ethics on its own. I find unhelpful pretensions that one can be fully explained or subsumed by the other. For natural historians are right to insist that humans as animals are subject to causalities of which they are not aware. But the social historians are right to insist that self-awareness and purposes matter. To repeat, we cannot step directly from the one to the other. This book follows them into the middle ground of social interaction, where people are provoked to cooperate or dispute, to explain themselves to one another, and above all, to see themselves through one another's eyes—or refuse to do so. If we are to grasp ethical life as something both natural and social in character, both innate and historical in its origins, we might start by examining some of the points of articulation where natural and social history approach, as well as push back against, one another. That examination is what this book aims to accomplish.

SOME QUESTIONS ABOUT ETHICAL LIFE

What are the stakes in raising such questions at all? Before I proceed, let me be clear that saying ethics is a ubiquitous feature of human life does not mean that all people are inclined to the good, an assertion so obviously absurd that it's hardly worth denying. Perhaps less obvious is this: I do not mean that even good people are likely to come to a consensus about what ethics entails. This claim requires more demonstration, on which more below. For now, it is enough to observe that the ubiquity of ethics offers no guarantees: people can assert diametrically opposed positions or values, such as hierarchy and equality, loyalty and justice, or fairness and discrimination, with equal ethical conviction. Rather, this book starts with the proposition that, with some borderline exceptions such as psychopathology, humans are the kind of creatures that are prone to evaluate themselves, others, and their circumstances. They may act in defiance of those evaluations but are rarely just indifferent to them. Consider the following stories, each of which exemplifies some of the problems with which research in ethics is grappling. The first and third are famous thought experiments; the rest are actual events.

The first story, known as the "trolley problem," has given rise to an enormous amount of discussion among philosophers and psychologists (for the original versions, see Foot 1967 and Thomson 1976; a recent popularizing summary is Edmonds 2013). Its basic form presents you with two imaginary scenarios. A runaway trolley is hurtling down the tracks at a group of five people, who will be killed if you don't intervene. In one scenario, you can pull a switch that diverts the trolley onto another track, where it will hit only one person. Utilitarian reason says that the death of one person is better than that of five. Most people who are presented with this situation in experimental settings agree and say they would pull the switch. The interesting complication arises in the second scenario. The five people are at risk as before. Now there is a man standing on a bridge over the tracks. He is so fat that were you to push him off the bridge, his body would stop the trolley. The utilitarian calculation remains the same: save five lives at the cost of one. But it turns out that most people balk at the idea of pushing the man to his death.

I will not reproduce the various attempts to explain the differences between the two responses and the endless variations they have given rise to. We will return to some of these topics in the next chapter. Here I want to make just a few observations to clarify the approach to ethics taken in this book. Obviously the trolley scenario is highly artificial, although analogous problems do arise, for example, in warfare and medical triage. Moreover, as historians and anthropologists will quickly note, the results are assumed to apply to all humans, yet the subjects of such experiments are usually drawn from a much narrower range, typically educated members of present-day urbanized, industrialized societies—serious problems arise when you try to set up the problem in other cultural contexts (Bloch 2012: 65–66). Still, the findings are provocative. What is more relevant for the purposes of this chapter, however, is the way in which the trolley problem depicts "ethics." The ethical problem is presented as a discrete event that requires a single decision and transpires within a brief time frame. That decision is taken by a lone individual who contemplates a limited set of clear options, which have immediate and unambiguous results. Those results can be measured on a single scale of value, numbers of lives saved. The experiment takes its interest from the contrast between ideal and actual responses to the emergency. The ideal is based on the assumption that there is a rational solution revealed in the consequences of each choice; the discussion is provoked by the ways people's actual gut feelings deviate from that solution. In short, the time frame is narrow, the social focus is on the individual actor, and the basic contrast is between rational and irrational decisions. Some aspects of ethical life are like this, but much is not.

Here is another story about a momentary decision, which opens up the range of questions we might need to take into account. It concerns a friend of mine, whom I will call Sally. Sally is a social worker in her fifties, married to a physical therapist. They have one grown child and another who still lives at home. They get by, but their financial situation is neither easy nor secure. Sally is the main breadwinner in the family, since her husband has been unable to find full-time employment in recent years, due to government budget cuts. For the last decade or so, Sally worked for an adoption agency run by a religious organization. This organization has never accepted unions between homosexuals and has a clear policy of refusing to help gay couples adopt children. One day Sally decided that in good conscience, she could no longer work for an agency that held such a policy and abruptly, and without consulting her husband or children, quit her job. She felt that she simply couldn't live with herself otherwise. She had nothing else lined up and in the year or so since has been semiemployed like her husband. Needless to say, this has rendered the family finances even more uncertain.

Now here are some ways we could tell this story. It shows that people are not driven only by egocentric calculations of gain. Ethics, in this perspective, stands in opposition to the values of economic rationality and to the idea that people's motives are always selfish. (But then the same can be said of the religious morality that leads the agency to reject gay applicants.) It also stands for the role that abstract, general ideas, such as justice or equality, might play in specific, concrete actions, such as quitting a job, and in more general dispositions, such as one's politics. At the same time, the thought that she could not live with herself otherwise reflects Sally's stance toward her own life, not just toward gay couples. And it shows someone who was willing to put her immediate family at risk (something that could be construed as unethical) for the sake of people known to her only as members of a general social category (gay couples)—that is, someone whose moral circle has expanded from the narrow confines of those closest to her. The story could also be represented as a narrative of ethical progress. We might imagine Sally acting quite differently a generation ago. Even ten years ago she worked for this agency with few qualms. The rise of gay marriage as a civil rights cause, along with its extraordinarily rapid acceptance in the United States, has been a remarkable social transformation. So if ethics is supposed to be solid bedrock, how could that happen? Yet another thing: Sally put her own family at risk. What ethical calculus allows her to treat their interests as less important than those of unknown strangers? A utilitarian might say that she was right to sacrifice a few individuals for a greater good; a certain kind of traditionalist might say that the obligation to kin is primary; and a virtue ethicist might go either way, depending on what Sally's actions say about her character.

Both Sally's choice and the trolley problem bear echoes of the conundrum posed by the English thinker William Godwin in the eighteenth century: If a house is burning, and I can save either Bishop Fenelon (an important social reformer and defender of human rights) or his chambermaid, but not both, which should I save? Godwin gives an early version of what would become a utilitarian answer. The rational choice is that which results in the greater good overall:

> Supposing the chambermaid had been my wife, my mother or my benefactor. This would not alter the truth of the proposition. The life of Fenelon would still be more valuable than that of the chambermaid; and justice, pure, unadulterated justice, would still have preferred that which was most valuable. (Godwin 1793: 83)

Accordingly, the bishop should be saved because his life has greater social value than the chambermaid's. But what if the maid is also my mother? Should calculations of utility trump the ethics of kinship? Godwin thinks so. But if they do, what kind of person would that show you to be? As the philosopher Bernard Williams remarks, if you hesitate in order to work out the justification for saving your mother, rather than instinctively pulling her from the flames, that is "one thought too many" (1981: 18). Considerations like these ask us to shift our attention from decisions to personal character and from the individual at one moment to his or her social ties to others over the long run.[1]

[1]It matters where utilitarian calculation starts. In *The Theory of Good and Evil*, a sober Oxford don writes, "The lower Well-being . . . of countless Chinamen or negroes must be sacrificed that a higher life may be possible for a much smaller number of white men" (Rashdall

These are normative questions, concerning what one *ought* to do. But as an empirical problem, how do we understand what Sally *actually* did? To understand her decision, do we look to psychology? Politics? Religion? And must we seek ethical heroes for counterarguments to self-interest? Heroes are few and far between: How will they help us understand the ethics that runs quietly through ordinary everyday activities, what I am calling "ethical life"?

One way to respond to such questions is to ask how local cultures shape the ethical choices and values of ordinary people. Here's a story from my own fieldwork in the 1980s and 1990s, on the island of Sumba, a rural Indonesian backwater (Keane 1997). Unlike most Indonesians, Sumbanese never converted to Islam, and until fairly recently they had limited contact with the dominant ethnic groups in the archipelago, their Dutch colonizers, and the nation-state that succeeded them. Much of Sumbanese life at this time was oriented around a relatively self-contained set of local values (but see Keane 2007). These values played into one of the key structural features of Sumbanese society, something anthropologists call asymmetrical marriage alliance. Sumbanese are born into their father's clan. Each clan is allied with certain other clans through marriages. In each generation, new marriages should renew those alliances. The way this works in practice is that a man is supposed to marry a woman from the same clan that his mother came from. The ideal marriage, because it is the closest way to reproduce his father's marriage, is for a man to marry his mother's brother's daughter (thus a woman should marry her father's sister's son). These alliances are asymmetrical: the worst thing a man could do is reverse the directions and marry a woman from the clan into which his sister should marry. Although clans are large enough, and the ways one defines kin are flexible enough, that there is some room for individual choice, alliances are a matter of collective interest and are negotiated by teams of elders from the clans involved. Marriage is far too important to be left to the personal preferences of the future husband and wife. It is

---

1907: 238–39; I thank Elizabeth Anderson for the reference). The point is not that utilitarianism is conducive to racism—historically the contrary was the case—but, rather, that the logic of the calculus does not guarantee the rationality of its premises.

also too expensive for any individual to sponsor, since the alliance is established through the elaborate negotiation and exchange of valuables such as pigs, horses, gold, and ivory, which reinforce ongoing relations of reciprocity and debt between affines. These negotiations and exchanges provide a public stage on which clans display their status; elders, their political clout; negotiators, their command of poetic speech; and individuals, their wealth.

Many Americans to whom I have described the Sumbanese marriage system react strongly. It runs against some of their core ethical values, such as individual autonomy, the free choice of a spouse, the idea of a love match and companionate union, and the elevation of sentiment over material goods in family life. It is against this background that I had a conversation with the elderly mother in the family with whom I lived during my fieldwork. Having talked endlessly about their own marriage system, she asked me whom my people are supposed to wed and what goods we use to accomplish it. When I told her that it is up to the individuals themselves, that there are no rules except for the prohibition on incest, and that we do not give goods in order to do it, she was visibly appalled. Thinking about my reply for a moment, she finally exclaimed, with shock, "So Americans just mate like animals!"

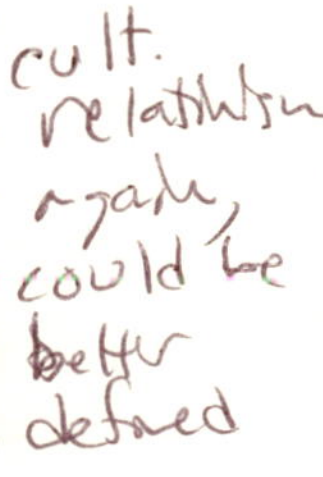

A conventional way to tell this story is as an illustration of cultural relativism: they have their values, and we have ours, and neither should be judged in light of the other. The clash between the two value systems has the salutary effect of denaturalizing what had seemed natural and fundamental to the naive person on either side. From this denaturalizing effect, one might then draw the conclusion that values are social constructions, each system wholly distinct from, or even incommensurate with, the other (Povinelli 2001). But the idea of cultural relativism has not always fared well, even among anthropologists. For one thing, the idea that cultures are more or less bounded entities, self-contained and internally consistent, has been hard to sustain in a world of constant migration, state penetration, mass media, global religions, and so forth (Appadurai 1996; Gupta and Ferguson 1997). A veiled Muslim woman who is the paragon of virtue in Algeria might find herself the object of moral indignation in France; so too the scantily clad German tourist in Java. Nor are cultural complexity and permeability necessarily just modern phenomena: some would argue that cultural worlds have always been exposed both to "external" influence and to "internal" contradictions by their very nature (Appadurai 1996; Clifford and Marcus 1986; Marcus and Fischer 1986; Rosaldo 1989).

Here is another angle: the ethics underlying my Sumbanese friend's reaction is not entirely unrecognizable even to a freedom-loving American. Although the values in each marriage system seem directly opposed to one another, this woman appeals to some other principles that look familiar. She recognizes that different communities have different marriage systems. After all, that is why she asked me the question. What makes the Sumbanese version distinctly ethical is, in part, the way in which it imposes external obligations and constraints on individual actors, in the name of some larger social good. Sumbanese are well aware that one might yearn to marry someone against the rules—and sometimes people do, although at considerable social cost. Moreover, they tell myths about ancestors whose supernatural powers included the ability to marry without marriage payments, stories whose appeal to listeners hints of wish fulfillment. So the sense of constraint is real and is linked to the sense of being ethical. It limits one's own willfulness. Those limits take concrete form not just in rules but in social interactions with other persons, who matter to one's own self-esteem. That very sense of limitations suggests yet another facet, that to be ethical is to be invested in a way of life and to live up to some vision of what a good person ought to be. Finally, an American might also recognize this aspect of my friend's remark: being ethical makes you human. To act without restraint is to be an animal.

Cultural accounts have their limits. People contradict one another, and individuals themselves are inconsistent, to say nothing of self-deceiving, so whom do we believe? And some ethical insights are innovative or idiosyncratic by local standards. Here's one example. During World War II a Polish peasant woman happened to overhear a group of her fellow villagers propose throwing a little Jewish girl into a well. The woman said, "She's not a dog after all," and the girl's life was saved (Gilbert 2003: xvi–xvii). To a philosopher, what might be striking here is the absence of principled justification or indeed any serious moral argument at all (Appiah 2008: 160). We may wonder how much conscious ethical reflection this woman's quip required on her part or on that of the people she addressed. It seems that she merely invoked, in a rather off-the-cuff way, a commonsense category, which reframed the situation so that the others could see what they had proposed in a new light. To some philosophers, this apparent lack of reflexivity may cast doubt on exactly how we should count this as a full-fledged ethical act.

An alternative approach would place the act in its cultural context. Although we may conclude that the Polish woman drew on a local category, clearly it was not until that moment salient to those who had, perhaps, taken the girl to be some kind of vermin. There is no reason to think that this woman did not share all the usual background beliefs and values with her fellow villagers: in this case, the explanatory power of "culture" alone doesn't seem to get us very far. But neither does innate human psychology, for the same reason, since it should apply equally to that woman and to the other villagers. Moreover, against the cheerful claim that this woman's instincts reveal a bedrock humane intuition, perhaps offering a clue to some universal basis for virtue, we would need to recall that a similar sort of gut reaction can find differences of skin color, sexual orientation, religion, dress, or eating habits immoral, fundamentally repugnant, and even inhuman (Haidt 2001; Rozin and Nemeroff 1990; Rozin and Royzman 2001).

The Polish villager's intervention points to some key questions for any empirical research into ethics: What are the relations between her gut-level response, on the one hand, and explicit modes of argument and reasoning, on the other? How does either of those articulate with taken-for-granted community norms and habits and their histories? Does a naturalistic explanation of that gut-level response—perhaps in affective, cognitive, or neurological terms—have any bearing on what happens when people appeal to norms, reason with one another, fault others, or justify themselves? Or vice versa? What made this Polish villager's intervention work? What gave her a voice in this situation, when we might imagine that some other person would have gone unheeded? How do we evaluate its success within the larger context of ethical failure surrounding it?

The Polish woman wins the day by invoking the ethical implications of an ontological category, with an implicit syllogism: because the girl is not a dog but a human, therefore she is owed what we owe to a human. But once we bring in ontology—those background assumptions about reality that are implicit in a certain way of life—we find ourselves back at the problem of relativism again.[2] For not everyone agrees on all the same ontological premises. Communities that agree on most aspects of reality (fires need dry kindling, crops need water) may differ vastly in how they answer the question "What can count as an ethical actor?" In the contemporary West the ethically responsible self is usually—but not always—considered to be bounded by birth (or maturity) at one end and death at the other. Not so in the various South Asian theories of karma, based as they are on the doctrine of endless cycles of rebirth; they teach that individuals suffer the consequences in this life for misdeeds they performed in previous lives that they cannot recall but for which they remain, in some sense, responsible (Babb 1983; Doniger 1980; Fuller 2004). Nor does responsibility necessarily stop with humans. Herodotus (1997: 525) reports that Xerxes had the Hellespont whipped and verbally chastised for destroying a bridge; medieval European courts punished animals for crimes (Evans 1906). One need not venture so far: present-day middle-class Americans differ among themselves over such basic questions as the existence of angels, the reality of the immortal soul, the personhood of the fetus, the intervention of God in one's personal life, the responsibilities of corporations, and the rights of animals.

Listen to ethnographer Paul Nadasdy recount his experience of learning to hunt with Kluane people in the Yukon:

> The first time I found a live rabbit in a snare was something of a crisis. I was alone, and I knew I had to break its neck. Never having killed anything with my bare hands before, I was not really sure what I was

[2] I use *ontology* in the anthropological, rather than philosophical, sense (see Descola 2013; Povinelli 2001; Sahlins 1985; Vivieros de Castro 1998; as well as discussion in Keane 2013). It does not refer to ultimate constituents of reality but, rather, to assumptions that guide people's observable actions. Some of these are presumably universal. Others are more local: if the lightning that struck my house is just meteorological bad luck, it makes no sense to ask who is at fault; on the other hand, if it was due to spirits, I had better find out what angered them.

> doing. The animal suffered as a result, and I felt terrible. . . . The next day, . . . I told Joe Johnson [a Kluane elder] . . . how badly I felt about the rabbit's suffering. He told me that I must never think that way. The proper reaction, he said, is simply to say a prayer of thanks to the animal; it is disrespectful to think about an animal's suffering when you kill it. I did not understand that at first. A couple of months later, however, Agnes Johnson . . . told me that it was "like at a potlatch." If someone gives you a gift at a potlatch, it is disrespectful to say or even think anything bad about the gift or to imply that there is some reason why they should not have given it to you. . . . It is the same with animals, she said. If they give themselves to you, you say a prayer of thanks and accept the gift of meat you have been given. To think about the animals' suffering, she said, is to find fault with the gift, to cast doubt on whether the animal should have given itself to you in the first place. To do this is to run the risk of giving offense and never receiving such a gift again. (2007: 27)

Kluane hunters, in other words, take their prey to be persons with whom they enter into social relationships guided by the ethics of reciprocity. That basic ethics of reciprocity in itself might not look so unfamiliar to, say, urban Euro-Americans. The difference, of course, lies in the scope of appropriate ethical concern.

Similar statements about the personhood of animals and other nonhumans abound in the ethnographic record. When people talk like this, however, they are usually not just engaging in dispassionate metaphysical speculation (Keane 2013). Often enough, what is at issue is how one should properly interact with other beings. Anthropologist Irving Hallowell (1960) observed that the Canadian Ojibwa in the mid-twentieth century did not normally see important events as resulting from neutral causes. Rather, they were the result of acts carried out by some kind of person, which might be an animal or a human spirit. The ethical implications of this kind of ontology were spelled out by Knud Rasmussen, the explorer-ethnographer, who wrote of Arctic hunters such as the Inuit that "the greatest peril of life lies in the fact that human food consists entirely of souls" (1929: 56). When ontological assumptions differ, they may shape what kinds of entities should be objects of ethical concern and what kinds of beings can be held morally responsible for events.

Let me quickly point out two things about these statements. First, Kluane, Ojibwa, and Inuit are skillful hunters and observant naturalists who certainly do not rely just on prayers, magic, or gifts to obtain meat. Second, they are hardly unaware that humans and animals are different: as Nadasdy points out, no one sets snares to trap people and eat them. So what are we to make of such statements? This is hardly a settled matter among the ethnographers. But even a reader who finds it hard to imagine that a rabbit can really be an exchange partner who willingly gives itself up to the hunter might yet recognize the ethical obligations that Nadasdy's friend Agnes Johnson was talking about. Gift, reciprocity, and words of thanks might be applied to surprising social partners, but the ethical nature of the relationship that these acts invoke should not seem utterly unfamiliar. In the midst of alien ontologies, do we see the dim outlines of recognizable ethical intuitions? Is ethical concern something we can recognize even when applied to entities we might consider out of bounds? This book makes an argument that in many respects the answer will be a cautious yes and that to make sense of why that is so, we cannot rely on either psychological or cultural explanations alone.

These six stories point to some of the key themes this book will address. Some of these themes—such as desire, emotions, and beliefs—are often treated as matters of individual psychology. Others, such as altruism, utility, reason, freedom, and the ethical distinction between human and nonhuman, seem to fall in the domain of philosophical or other normative enterprises. Still others, such as politics, values, and cultures, are usually viewed in terms of social institutions. And some, such as voice, can be hard to pigeonhole. One of the tasks this book undertakes is to tease out the interconnections within this sprawling and apparently heterogeneous list. To start, let us consider some key terms: *ethics*, *morality*, *reflexive awareness*, and *affordance*.

DEFINING ETHICS AND MORALITY

I first began thinking about the sciences of ethics and morality while trying to understand the conversion of Sumbanese ancestral ritualists to the Protestant Christianity brought to their Indonesian island by twentieth-century Dutch colonial missionaries (Keane 2007). One of the central challenges this situation presented was making sense of how Sumbanese were able to rethink and change ethical values that, on the face of it, should have been part of those background cultural and ontological assumptions that are so deep and so world-defining that they can be almost impossible to question. But in this context "ethics" and "morality" seemed to be relatively straightforward concepts. They were defined in terms of an institutionalized religion with an explicit moral code. Matters became more complicated, however, when I ventured into the less self-conscious domains of habitual activities and everyday social relations that some ethnographers have called the "ordinary" (Das 2007; Lambek 2010). As I use it, "ethical life" starts from that sheer everydayness, that mere fact, as anthropologist James Laidlaw puts it, that people "are evaluative" (2014: 3). But as I began to explore other work by social scientists, I discovered that there is no consistency in how they use the words *morality* and *ethics*, which are often treated as requiring no definition at all.

A glance over some of the major writings in the anthropology of ethics and morality illustrates the point. In his 1925 essay *The Gift*, Marcel Mauss (1990) never defines *morality*, but it is apparent that he has in mind those obligations between persons that constrain their self-interest. Within the different African contexts they study, T. O. Beidelman (1980) uses *morality* to refer to character traits, and Wendy James (1988), to that which maintains a person's health and balance in the face of evil forces. For K. E. Read (1955), *morality* refers to specific rules and judgments, while *ethics* consists of the underlying ideas about humans and their relationships on which those rules are based. Arthur Kleinman (1998) seems to reverse this distinction, using *morality* to refer to ultimate values and *ethics* to speak of the explicit principles propagated by elites. Finally, *ethics* often refers to the regulation of a profession, as in "scientific ethics" or "business ethics" (Meskell and Pels 2005).

In response to this inconsistency, I have found it useful to keep in mind a distinction articulated by the philosopher Bernard Williams (1985). Williams is critical of a dominant view in modern Western philosophy that emphasizes obligations and blame and assumes they must be based on a wholly consistent system of highly general principles that should apply to all people regardless of their identities or circumstances. This emphasis, which he calls "the morality system," obscures other crucial aspects of what he calls "ethics." Whereas morality deals with such questions as what one should do next, ethics concerns a manner of life—not momentary events but something that unfolds over the long term and is likely to vary according to one's circumstances. Viewed from this perspective, the trolley problem addresses an issue of morality, and the Kluane rabbit hunters, the nature of ethics. Ethics is thus less about decisions and the rules that should govern them than about virtues, which "involve characteristic patterns of desire and motivation" (Williams 1985: 9). (Some psychological research has been taken to challenge the realism of this view of the virtues, but that discussion must wait until the next chapter.) Although both ethics and morality say something about what one owes to other people and how one should treat them, they differ in how they portray social relations. Many of the most powerful rules and obligations of the morality system are meant to be universal in application, drawing on principles that transcend any particular context or person, like Kant's Categorical Imperative. Moral obligations are the sort of things you might contemplate on your own. By contrast, ethics captures the way in which

> the agent's conclusions will not usually be solitary or unsupported, because they are part of an ethical life that is to an important degree shared with others. In this respect, the morality system ... conceals the dimension in which ethical life lies outside the individual. (Williams 1985: 191)[3]

This emphasis on the social nature of ethics is one reason why Williams's distinction between the two terms has been especially congenial to researchers working in historically and sociologically complex

[3]A parallel contrast exists in research in psychology. According to Lawrence Kohlberg (1981), child development should result in the acquisition of a mature sense of justice as something that is context-free and universal. Opposing this, Carol Gilligan (1982) argues that since people first are children of specific mothers, raised within networks of care, one cannot know what is good for someone in the abstract but, rather, only in particular social contexts. The philosopher Seyla Benhabib (1992) has made a cogent argument for reconciling the two into a more comprehensive perspective.

situations. It attends less to how ethics constrains people than to the ways it facilitates their ability to act and provides them with goals (Faubion 2011; Humphrey 1997; Laidlaw 2014).

We should not draw the distinction between ethics and morality so sharply that we are forced to exclude some of the phenomena we want to understand. As I read Williams, ethics does include the morality system—morality is just a special *kind* of ethics. It conceals but does not eliminate the ways ethics is socially embedded. And the ethnographic and historical records are indeed full of rules and obligations, put in very general terms, which are meant to be internally consistent, like the morality system Williams criticizes. Since these extend far beyond the tradition in Western philosophy that Williams had in mind, I will use the expression in the plural and propose that there are *many* morality systems, of which the tradition Williams attacks is only one example. In certain communities, following rules *is* what the virtuous life consists in. Here we might include my Sumbanese mother's view of the morality of kinship and marriage, which includes adherence to explicit sets of obligations and prohibitions, or the Hopi, who by one account treat ethical questions as concerning duties based on moral facts that one should know (Brandt 1954: 82). Other examples include imperial China and premodern Europe, where morality was often treated as something people could not be expected to grasp unless they had been instructed by authorities (Brokaw 1991; Schneewind 1998).

What often links rules and the virtuous life is reference to a deity. Sumbanese marriage rules, for example, are enforced not only by social means but also by the threat of sanctions from the spirits, which might take form as infertility, lightning strikes, or drought. More generally, the coherence and explicitness of religious morality systems are accounted for by their divine origins—their authority by the existence of a transcendental judge. For many secular philosophers, this disqualifies such systems from serious consideration.[4] Not so for the historian or anthropologist, since most of the people they

[4]Williams claims that the legalistic nature of the morality system is "modelled on the prerogatives of a Pelagian God" (1985: 38), echoing the position of Elizabeth Anscombe, that reducing ethics to obligation is incoherent "unless you believe in God as a law-giver" (1958: 6).

study have precisely such a view of the world. As we will see in chapter 6, some of the most historically influential morality systems are organized around the cultivation of piety. If Williams is right to insist we not reduce ethics to a morality system, we should still recognize that the production and inculcation of morality systems are among the looming historical realities we need to understand. Putting morality systems in the context of ethics encourages us not to take their existence for granted. Instead, we can ask what circumstances tend to foster or induce the development of morality systems: more or less context-free, more or less explicit, systems of obligations. This is the problem that this book takes on in part 3.

"Morality" can thus be treated as a special case within ethics. Studies that focus on virtues, values, and ways of life (like the values embodied in Sumbanese marriages) tend to fall under the rubric of ethics. Those that focus on obligations, prohibitions, general principles, systematicity, and momentary decisions (like the trolley problem) are treated as morality. But there is a great deal of overlap and interaction between these. Sumbanese social values and Kluane relations to animals do make reference to rules and obligations. Your resistance to pushing the fat man in front of the trolley may be due to what kind of person you want to be. I have found that in many actual instances, it is an artificial matter to try to keep the two distinct, and I have varied my usage accordingly.[5]

In this book, I will treat "ethics" as the more encompassing category of the two. The meaning of the word *ethics* as I use it here is very broad. It is tempting to follow U.S. Supreme Court Justice Stewart's famous definition of pornography, "I know it when I see it," or the words of the philosopher David Velleman, who says that since moralities are variations on themes that bear a family resemblance, "I do not offer a definition of what I mean by 'morality' or 'moralities.' I mean *that* family (you know which one it is)" (2013: 3). But this is unlikely to satisfy most readers. As a rough heuristic, I take ethics to center on the question of how one should live and what kind of person one should be. This encompasses both one's relations

[5] I thank Ian Hacking for permission to relax this verbal distinction. Commenting on an earlier essay (Keane 2010), he noticed that efforts to keep the terms separate could lead to contortions of prose that suggested they were squeezing the usage into artificial straitjackets.

to others and decisions about right and wrong acts. The sense of "should" directs attention to values, meaning things that are taken by the actor to be good in their own right rather than as means to some other ends. This refers to the point where the justifications for actions or ways of living stop, having run up against what seems self-evident—or just an inexplicable gut feeling. As such, values can also motivate the sense that the rules and obligations of a morality system are binding on one's specific actions. For even the taboo whose justification is simply that it was dictated by the ancestors can be understood this way, since as those who observe the taboo see things, it is not necessarily a means to some further end (Valeri 2000).

One way to grasp the link between values and how one should live has been summarized by the philosopher Elizabeth Anderson this way: "Value judgments commit one to certain forms of self-assessment" (1993: 3). That is, there is a crucial link between one's sense of self-worth and what one values beyond the self. Anderson goes on to say that because the meaning that values hold is public, one's sense of self-worth is something that others can grasp as well. Indeed, much of the empirical evidence that we will examine in the following chapters concerns how people evaluate one another and how that mutual evaluation in turn reflects back on each one's self-understanding. To invoke Velleman again, a core element of ethics (or what, reflecting the unruly application of these terms, he calls morality) is "valuing the personhood of people" (2013: 72). One of the challenges this book takes up is to justify this claim on empirical grounds and give it some psychological, ethnographic, and historical specificity. It aims to do so not just in the traditional anthropological manner, by demonstrating that cultural worlds vary, but also by exploring different scales of inquiry, including the budding abilities of young infants, the routines of conversational interaction among adults, and purposeful large-scale social movements that take generations to unfold.

AWARENESS AND REFLEXIVITY

Cutting across the distinction between ethics and morality is another one, that between the tacit and the explicit, those background assumptions, values, and motives that go without saying or are difficult to put into words, on the one hand, and those that easily lend themselves to conscious reflection, on the other. This distinction does not map directly onto that between ethics and morality. Ethical life often involves psychological phenomena that work beneath the level of awareness, like one's emotionally powerful repugnance at pushing the fat man in front of the trolley (Greene et al. 2001). As we will see in the next chapter, people's gut-level responses to situations like that might, if they were asked to reflect on it, just induce what the psychologist Jonathan Haidt (2001) calls "moral dumbfounding," a puzzled inability to give good reasons in support of a strong ethical intuition. Ethical life also draws on social and cultural background assumptions, like Kluane ideas about the personhood of rabbits or Sumbanese assumptions about marriage. Although these assumptions can be made explicit, most of the time they are likely to remain unspoken—until someone like a moral reformer or an anthropologist asks questions about them. When either background assumptions or gut-level responses are put into words, they undergo changes in both their cognitive and their sociological character. As a result, verbal report is at best a poor guide to the sources of people's feelings and decisions or even to what they know or believe (see Bloch 2012 for an overview). But ideas and values that are subject to conscious apprehension can have important social and historical roles. For one thing, they are more easily transmitted to distant times and places, for instance, as doctrinal teachings and codes of conduct (Silverstein and Urban 1996). This is one reason why morality systems tend to favor explicit formulations. By the same token, they are also rendered easier to scrutinize from the outside, as it were, and so more subject to post hoc justifications, to criticism, and to instrumental manipulation. Indeed, several ethical traditions worry that self-consciousness will disrupt the spontaneity or disinterestedness that should mark virtue. According to Edward Slingerland (2007), a scholar of Chinese religion, early Confucian and Daoist philosophers grappled with the paradox that results from holding both that one should actively strive to be virtuous and that the purposeful effort contaminates the result. We will examine all these issues in more detail in the chapters that follow.

If we accept that morality systems and ethics can be treated within a single field of inquiry, then what should we make of the distinction between explicit and tacit, what is put into words and what remains taken for granted or beneath awareness altogether? We might divide the question into two parts: First, what conditions induce explicitness, and second, what are the practical or conceptual consequences of explicitness? To see what is at stake here, let's turn to another contrast. Many definitions of ethics in the Western philosophical tradition turn on a distinction between the causes of an action and the reasons for it (Darwall 1998). In these traditions, for an action to count as ethical it must be directed or justified in the light of some values recognized as ethical by the actor (Parfit 2011). This requires both some degree of autonomy from natural causality or social pressure (one could have done otherwise) and some quality of self-awareness (one must know what one is doing). Something like this distinction apparently holds even in traditions as far from Western philosophy as South Asian karma. At first glance it may seem mere fatalism to attribute my misfortunes to actions carried out in a previous life that I cannot remember. But in some common views of karma those actions are *ethical* misdeeds because they were carried out by those who were responsible precisely because, at the time of the misdeeds, they had volition and knew their moral obligations (Babb 1983).

Even the social theorist Michel Foucault (1985, 1997), an heir to Nietzsche's skeptical quarrel with much of the Western philosophical tradition, holds that ethics depends on reflexivity. In Foucault's view, this reflexivity turns on a capacity for self-distancing, since "thought . . . is what allows one to step back . . . to present [one's conduct] to oneself as an object of thought and to question it as to its meaning, its conditions, and its goals" (1997: 117). This takes the relative freedom or autonomy that defines an action or stance as being ethical to be inseparable from heightened self-consciousness (Schneewind 1998). Foucault, in this respect at least, seems to be working within the broad parameters of that tradition that places ethical life in the domain of reasons and justifications.

Challenging this tradition are the apparently corrosive effects of both the natural and the social sciences on Euro-American ethical thought. Since the era of Darwin, Marx, Comte, Quetelet, and Freud, both naturalistic and sociological explanations have challenged the human self-mastery and self-awareness implicit in the morality system. By pointing to forces and causes beyond ordinary awareness, these explanations can seem to debunk the feeling that your actions are guided by your own conscious purposes. The neurologist and "new atheist" Sam Harris (2012) gives one example. In 2007, two men in Connecticut committed a completely unmotivated rape, murder, and arson. It turned out that they suffered from brain malformations that deprived them of any capacity for empathy. Harris writes, "Whatever their conscious motives, these men cannot know why they are as they are. Nor can we account for why we are not like them" (2012). In his view, the third-person perspective that reveals mechanical causality simply trumps the first-person point of view, the actor's own grasp of what he or she is doing. Harris asserts that such findings eliminate any role for the concepts of morality or justice. Coming from a very different intellectual tradition, heading toward different conclusions, the sociologist Zygmunt Bauman (1988) notes a parallel implication. To see human activity as the product of ideological state apparatuses or neoliberal economics is a "science of unfreedom" (see Laidlaw 2014). As with neuroscience, so too sociology: causal explanations that cast doubt on freedom likewise seem to eliminate responsibility. This is exactly what the hoods in the musical *West Side Story* try to take advantage of when they address a policeman: "Dear kindly Sergeant Krupke, / You gotta understand, / It's just our bringin' up / That gets us out of hand. / Our mothers all are junkies, / Our fathers all are drunks. / Golly Moses, natcherly we're punks!" (Sondheim 1957). These approaches exemplify the problem faced by any concept of ethics that relies on notions of self-awareness, self-mastery, or freedom.

But if people are largely unaware of who they are and why they do what they do, we may ask with Harris or Bauman whether their characters or their actions can really count as ethical at all. What would distinguish ethics from matters of taste, operant conditioning, or obedience to authority? What would make an instinctive revulsion against pushing a fat man in front of a trolley part of the same family of considerations that includes support for gay marriage, respect for rabbits, rejection of ethnic cleansing, and obedience to ancestral marriage rules? The approach I take in this book is twofold. First, I argue that reflexivity is not a necessary precondition for ethics as such. But it can play a catalyzing role in producing that public knowledge that feeds back into people's unself-conscious responses to other people and their actions. For people's ethical intuitions may not always be subject to reflection—hence the common gut reaction against pushing the fat man in front of the trolley and perhaps the Polish woman's comment that saved the Jewish girl. However, in order to identify certain situations as posing a distinctively ethical question or an individual as having a character of a certain ethical kind, people can draw on those descriptions that are available to them. Those descriptions—some might be summed up in simple words such as *lie* or *loyalty*, others require more elaborate discussion—are public knowledge: you can expect other individuals to recognize them much as you do. In its fullest form, this public knowledge plays a crucial role in defining for people whether a given act or way of life *is or is not an ethical matter at all*. Second, I pay attention to the social circumstances that induce reflexivity. They are crucial to understanding ethics, because they also enter into the dynamics of recognition and self-recognition that underlie the sense of self-affirmation Anderson refers to and the valuing of personhood of which Velleman speaks.

In short, taken as an object of empirical research, ethics is defined neither by rationality nor by special kinds of self-consciousness. Nor should we decide in advance what, in any given empirical case, will turn out to count as ethical. Sally's stand in defense of gay adoption confronts opponents who may take their position to be just as firmly ethical in character. Yukon rabbits may seem off the radar altogether. But because, as I will argue, ethics draws on a *heterogeneous* set of psychological and sociological resources, some account is needed for what *groups them together* as ethical. As Velleman's invocation of the idea of family resemblance suggests, this grouping might not be due to any single essence that they all have in common. Certainly it does not depend on specific content. The ethnographic evidence makes clear that what counts as ethical in one social context—what one eats or how one dresses, for example—or who is the proper object of ethical concern—say, rabbits or ecosystems—lies altogether outside the domain of ethics in another (Shweder, Mahapatra, and Miller 1990). Given the heterogeneity of all the things that might fall under the rubric of "ethics," it is the existence of publicly known descriptions and categories and their role in people's own ability to reflect on themselves and their situations that help define the common threads of value running through them.

Any investigation of how the domain of the ethical comes to be defined needs to include—but not simply rest with—the dynamics of reflexivity. The evidence in the chapters that follow suggests that we should not put individual psychology, private contemplation, or cultural and religious systems at the center of that dynamic. Rather, in order to understand what produces ethical reflexivity, we must look at what happens when all of these are put into play in social interactions. For social interactions are the natural home of justifications, excuses, accusations, reasons, praise, blame, and all the other ways in which ethics comes to be made explicit. Put crudely, they always require a self and an other to whom that self owes an accounting. In part 2, we examine patterns of social interaction as critical components of ethical life. What's crucial here is not to take the domains of reflection and talk in isolation or to treat them as simply expressing preexisting cognitive or emotional dispositions, moral codes, ethical precepts, cultural values, or social categories. To understand how ethical reflections emerge, they must be situated in relation to other dimensions of ethical life. These include both those psychological processes that work beneath people's normal awareness and the historical ones that may range beyond it.

To summarize thus far: many traditions of moral thought propose that ethics must have a universal and comprehensible basis if it is to make serious claims on people. Empirical research has long posed two kinds of challenge to these assumptions. One is relativist: the historical sciences often stress the existence of dramatic cultural differences against claims about the universality of ethical intuitions. By contrast, naturalistic explanations in psychology or neuroscience often suggest that apparent diversity masks shared human traits. But such accounts pose another challenge, seeming to replace judgment with causality. As I have noted, this runs counter to one philosophical position, that ethics cannot just be doing the right thing but must be doing it for the right reason. If so, either causal explanations are not really about ethics or else they require that ethics be redefined.

How do we reconcile explanations that posit causes that people are not conscious of with the idea that ethics involves self-awareness? What place should cultural and historical differences have in our understanding of ethics as a dimension of all human communities? To address these questions, this book draws on research findings from across several disciplines, especially psychology, conversation analysis, ethnography, and social history. The purpose is to reconstruct an approach to ethics that looks at the points of articulation among these domains. It aims to illuminate the dialectic between the shared human capacities explored by fields such as psychology and the variability that is at the heart of ethnography and history. *Dialectic*, in this sense, is an imprecise term, meant only to indicate that the relations among these dimensions of human life are neither wholly deterministic nor unidirectional. Sometimes they have a character similar to what philosopher Ian Hacking calls looping effects: "People classified in a certain way tend to conform to or grow into the ways that they are described; but they also evolve in their own ways, so that the classifications and descriptions have to be constantly revised" (1995: 21). But looping does not seem to cover all cases. We also need a concept that will allow us to grant the reality of certain properties that humans possess, without forcing us to conclude that these properties necessarily *determine* the results in every case. Here we might speak of ethical affordances.

ETHICAL AFFORDANCES

By *ethical affordance* I mean any aspects of people's experiences and perceptions that they might draw on in the process of making ethical evaluations and decisions, whether consciously or not. The idea of affordance originated in the psychology of visual perception but has influenced wider developments in situated cognition (Clark 1997; Hutchins 1995) and cultural psychology (Wertsch 1998). As defined by psychologist James J. Gibson, "the affordance of anything is a specific combination of the properties of its substance and its surface" in light of what it offers, provides, or furnishes for the animal that perceives it (1977: 67–68). Or as the philosopher George Herbert Mead put it, "The chair invites us to sit down" (1962: 280). Gibson stresses that although the properties are objective phenomena, they serve as affordances only in particular combinations and relative to particular actors. Thus,

> if an object that rests on the ground, has a surface that is itself sufficiently rigid, level, flat, and extended, and if this surface is raised approximately at the height of the knees of the human biped, then it affords sitting-on. . . . [But] knee-high for a child is not the same as knee-high for an adult. (Gibson 1977: 68)

Two crucial points in this original definition are, first, that affordances are objective features in contingent combinations and, second, that they only exist as affordances relative to the properties of some *other* perceiving and acting entity. Gibson was interested in animals' relations to their environment, not just humans, so there is another implication as well. One's response to an affordance does not depend on cognitive representations. A weary hiker may ease him- or herself onto a rock ledge without conceiving of it as chair-like or even being aware that he or she is doing so at all. But the idea of affordance does give great weight to perceptual experience with the forms of things (Keane 2003).

Affordance is an alternative to a classic argument from design—that if something functions in a certain way, then that must be its original purpose. What is crucial here is the fact of (mere) potentiality: a chair may invite you to sit, but it does not *determine* that you will sit. You may instead use it as a stepladder, a desk, a paperweight, or a lion tamer's prop or to prop up an artwork, to burn as firewood, to block a door, or to hurl at someone. Or you may not use it at all. Affordances are properties of the chair vis-à-vis a particular human activity. As such they are real and exist in a world of natural causality (chairs can hold down loose papers or catch fire), but they do not induce people to respond to them in any particular way. I want to argue that this quality of potentiality is a necessary consideration in any empirical approach to ethics, if we accept two basic propositions: first, that ethics has some naturalistic components and, second, that to be properly ethical, an act or way of living cannot simply be the inevitable outcome of a set of causes.

The concept of affordance has been usefully extended since Gibson's original formulation. It helps industrial designers (e.g., Norman 2002) pay attention to how tools and other devices give their users cues to how they should be used. Archaeologists (e.g., Knappet 2005) have been interested in finding those cues in artifacts, as evidence for the concepts and purposes of their long-vanished users. And cultural anthropologists (e.g., Ingold 2000) have sought in affordances a way to understand things such as weavers' embodied knowledge of their craft and its raw materials. These various approaches have brought out three aspects of affordances. One is that affordances usually work along with other sources of information, such as cultural routines. Carl Knappett (2005) points out that a postbox has affordances very similar to a trash can, so some background knowledge is still needed to use each the way it was intended. A second point is that people seek out affordances not as a matter of contemplation but in the course of practical activity (Norman 2002). The nature of that activity and its goals will affect what affordances they discover in their surroundings. A chair is for sitting if that is what one is seeking, but given another activity, it is for blocking the door or holding down a windblown tarpaulin. The third is that for humans affordances can be social. According to the evolutionary anthropologist and psychologist Michael Tomasello (1999: 62–63), a critical juncture in a child's cognitive development comes with the ability to enter into joint attention to an object with another person. Mediating this social relationship is shared alignment to certain aspects of the object. Among mature adults, this alignment can become a matter of negotiation, disagreement, or exclusion. This is apparent, for instance, in the process by which police decide whether an object is evidence of criminal activity. Investigators

> evaluate the objects that have been seized from the persons accused, categorize them, and decide what they imply for their (ex-) owners. This occurs within an interaction context of multiple agents, such that the affordances of the objects are explored and shared. . . . Consequently the police officers go through an extensive shared manipulation of the knives; handling them, understanding their particular affordances. They discover that one of the knives is burnt at the blade,

> prompting them to accuse its owner of using it for hashish. Another blade is broken—the owner claims to have done it opening a tin—the officers suggest he broke it trying to force open a car door. (Knappett 2005: 47)

The idea of affordance usefully draws us away from treating material forms as wholly transparent in three ways. First, it shows the role of past experience: people respond not just to immediate percepts in isolation but to recognizable patterns over time. Second, affordance can be understood to refer not to objects or people who interact with them but to entire situations (Kirsh 1995). Finally, the idea of affordance brings out the fundamental sociality of those situations.

To extend the concept of affordance to ethics, we might start with the simple observation that not just a physical object but *anything at all* that people can experience, such as emotions, bodily movements, habitual practices, linguistic forms, laws, etiquette, or narratives, possesses an indefinite number of combinations of properties. Even the well-known psychological phenomenon of "hearing voices" can be taken up as an affordance, providing evidence of benevolent gods in Ghana, kindly relatives in India, or hostile strangers in the United States (Luhrmann et al. 2014). In any given circumstance, properties are available for being taken up in some way within a particular activity, while others will be ignored.

As we will see in the next chapter, many researchers have argued that the child's cognitive, emotional, and social development is not simply a matter of unfolding genetically preprogrammed potentials. The ability to acquire language, for example, requires that the child be engaged linguistically by other people before reaching a certain age. Otherwise that cognitive window of opportunity shuts forever. Many other aspects of the child's mature capacities also come to fruition only when prompted through interactions with other people or aspects of his or her surroundings. Like language, those interactions and surroundings will bear the mark of specific social histories. In some respects, this prompting in turn is prompted, as the child actively seeks it out. The developmental relationship between the child and his or her surroundings is not simply one of learning from others or just the expression of innate abilities: it is, at least in part, one of *discovering what they afford*. The specific activities that facilitate the child's development may reflect particular ways of life. Which affordances the child takes up may be shaped by that "training in everyday tasks whose successful fulfilment requires a practised ability to notice and to respond fluently to salient aspects of the environment" (Ingold 2000: 166–67).

This account of the child's active role in learning is one consideration that has led me to prefer the idea of "affordance," rather than, say, "precondition," as a useful way to grasp the process. To call something a precondition suggests that there is only one relevant outcome. Affordances leave things more open-ended—without, however, turning the infant into a tabula rasa. I argue that the idea of affordance does a better job of illuminating links between the particularities of social and historical circumstances and the universal capacities on which ethical responses draw than do the more traditional versions of cultural construction. This argument aims to open up a more productive relationship between disciplines that stress diversity and change on a historical scale, on the one hand, and those that stress universality and change on an evolutionary scale, on the other.

Now child development is only one part of the evidence we will examine. But it is an important starting point in our discussion, because so many strong claims about innate ethical universals have rested on this research. Affordance is a useful way to understand how the kinds of data found within the distinctive research traditions into humans' natural and social histories might be connected. It suggests a way to explore their connections without assuming that they must lead either to sheer determinism, on the one hand, or to pure self-invention, on the other.

Ethical affordance refers to the opportunities that any experiences might offer as people evaluate themselves, other persons, and their circumstances. What those experiences can be will vary widely. In the next chapter, we will look at some of the fundamental capacities and propensities that children develop that enter into the ethical life of adults, such as empathy, intention-seeking, sharing, helping, conformity, discrimination, norm-seeking, and norm-enforcing. Among these, very young children develop an ability to draw inferences from what they perceive in order to impute intentions, desires, and beliefs to other people. This is sometimes called "mind reading," although it always depends on perceptions—it is not telepathy. It is impossible to really learn to speak without this ability. Viewed from one perspective, mind reading is a precondition for speech. But, as we will see in chapter 3, mind reading, especially in intention-seeking, can also be a source of ethical affordances. I can only help you if I have some grasp of what you intend to do or what you desire. You can only insult me in certain ways if I understand that your words were intended to wound. In both situations, the ethical character of the action presupposes mind reading. This ethical dimension can itself become the focus of people's attention. One reason why the existence of other people's hidden intentions can be a source of special anxiety is precisely because they can be malevolent or benign. In some societies, there is a history of fascination with these questions, and mind reading has become the focus of enormous attention. It may well be, for instance, that the modern English novel, with its focus on subjective experience, arose when and where it did (and not at other times and places) in part because of this fascination among eighteenth-century readers (Hunt 2007; Zunshine 2006). In other societies such as the Korowai of West Papua, as we will see in chapter 3, the opposite has been the case, and people deny even having the ability to guess at intentions at all (Robbins and Rumsey 2008). Mind reading and its denial are ethically fraught. To elaborate or deny one's mind-reading capacities, and the impulses behind them, is to respond to an affordance that human social cognition offers. Ethical affordances are those features of human psychology, face-to-face interaction, and social institutions that can be taken up and elaborated within ethical projects. They are part of what makes it possible for ethics to be both a universal feature of human existence as an animal species and something that has a variable social history.

OVERVIEW OF THE BOOK

This book is divided into three parts, looking in turn at psychology, everyday interactions, and social movements. One difference among these fields is the role played by people's self-awareness. At the psychological level, ethical affordances involve processes that mostly (but not always) operate beneath the level of ordinary awareness, such as seeking out another person's intentions or grasping his or her emotional state. By contrast, social movements respond to moral systems that have been developed by highly self-conscious historical actors such as religious reformers or lawmakers. Depending on which dimensions of ethics they look at, and at what scale, the natural and social sciences will see different processes at work. As the anthropologist James Laidlaw points out, "Moral codes and ethics must be distinguished analytically, because they may change independently" (2014: 111). One reason for this is that they tend to draw on different kinds of affordances. Some, for example, derive from the mechanics of human neurophysiology; some, from the dynamics of face-to-face interaction; some, from the logic of arguments; and some, from the power of institutions. A central argument of this book is that what links the psychological and historical dimensions of ethical life is the dynamic of everyday social interaction. Psychological processes become visible and ethically pertinent when other people have to respond to them. Explicit moral reasoning and justification arise most often when people are called on to account for themselves to others. This is not to deny the creative and critical powers of introspection, but, as I will suggest in the chapters that follow, even solitary thought draws on the resources of a life filled with actual conversations and imagined interlocutors. When moral ideas become explicit, they are made more readily available for purposeful historical actors to work with. Social interactions, in turn, must draw on the resources of psychology and ethical history. Each of these dimensions of ethical life serves as a context for the realization of the others.

This book does not try to encompass the full range of current naturalistic explanations for ethics but concentrates on the borders where natural and social sciences come closest, especially those points where they seem to be looking at the same phenomena. Still, the reader may want to know where this book does or does not touch on evolutionary theory, given its dominance in the natural sciences. Can evolution explain what makes humans the kind of animal that is capable of taking, and perhaps compelled to take, an ethical stance toward other humans and their actions? Evolutionary theorists sometimes put the question this way: If we take as our starting point the reproductive success of an entity such as a gene, or an individual organism, how do we explain altruism, defined as "behavior that benefits another organism, not closely related, while being apparently detrimental to the organism performing the behavior, benefit and detriment being defined in terms of contribution to inclusive fitness" (Trivers 1971: 35; see Wilson 1976)? We will return to the problem of defining altruism and its connection to ethics in the next chapter. For the moment, consider two basic approaches to this question, one looking to the logic of natural selection and another, to organic mechanisms. Although in principle they should converge, the research tends to work along separate lines. For example, to answer the question of "how cooperation can emerge among egoists without central authority" (1984: viii) the political scientist Robert Axelrod turned to game theory. His computer simulations showed that the most effective long-run strategies for the egoists led to cooperation. Since it is the logic of the system that produces the results, the model is meant to apply equally to microorganisms and political coalitions. Like artificial intelligence modeling of moral cognition (e.g., Wallach, Franklin, and Allen 2009), the system does not require any particular material processes.

By contrast, neuroscience looks for organic mechanisms to explain a rather heterogeneous set of definitions of *ethics* and *morality*, such as reciprocity, empathy, extended attachment, fairness, in-group identification, and disgust, not all of which map easily onto altruism. For example, neuroscientist Donald Pfaff (2007) argues that an impulsive and potentially self-sacrificial act, such as leaping onto a train track to rescue someone, is due to a momentary cognitive loss of information about self-identity. He proposes that this loss leads one individual to momentarily identify with another one, motivating him or her to prevent harm to that person. Interestingly, he finds more than one possible neuroanatomical explanation for this effect. Other researchers (e.g., Immordino-Yang et al. 2009) propose that the kinds of social effects on which phenomena we might call "ethics" depend arise from interactions between brain regions that in themselves serve quite different functions from anything particularly social. In fact, a many-to-one relationship between mechanism and function and vice versa seems to run through both the approaches I have sketched here. As the neuroscientist Valerie Stone (2006) remarks, brain functions that enable social cognition also serve a host of other purposes such as syntax, planning, and episodic memory: there is no firm evidence that they evolved to serve any one particular function. Such findings suggest why affordance can be a useful heuristic. It offers us one way of understanding how the results of natural selection might come to serve an array of different human purposes without reducing those purposes to a single deterministic source—while at the same time keeping their basis in the natural world clearly visible. More generally, if a given mechanism may serve many functions, and many mechanisms can serve the same function, explanations of ethics that rely on natural selection may turn out to give us only very general claims. For the purposes of this book I will bracket the question of origins—how people came to be the kind of animals capable of taking ethical stances—and treat as given the fact that this is how they have ended up. We can then concentrate on how natural processes articulate with the particular purposes and projects that loom large in people's awareness of themselves as agents who are guided by values and judgments—how those processes enter into diverse social histories.

The chapter following this introduction reviews some of the major findings in developmental, cognitive, and moral psychology that have been taken as evidence for the foundations of ethics. It looks at research on human capacities and propensities for things such as sharing and cooperation, intention-seeking, empathy, self-consciousness, norm-seeking and enforcement, discrimination, and role-swapping. It argues that they are necessary but not sufficient conditions for ethical life. What they help explain is what it is about humans that makes them prone to taking an ethical stance. But to understand how that mere potential can be realized in actuality requires attention to interaction. A central thesis of this section is that for the psychology of ethics to have a full social existence, it must be manifest in ways that are *taken to be* ethical by someone. Ethics must be embodied in certain palpable media such as words or deeds or bodily habits. The ethical implications must be at least potentially recognizable to other people.

To supplement what psychology shows us, we need to put it in the context of an account of the self in relation to other people. This is the focus of the second set of chapters, on social interaction and intersubjectivity. Chapter 2 works at the intersection between psychology and the study of conversational interaction. The ethical implications of the basic features of interaction are registered in the ways people probe one another's intentions and character, for example, or take others to be according or denying them recognition. Picking up on these themes, chapter 3 looks at a variety of ethnographic cases to show how recognition and intentionality are elaborated and brought into focus in different cultural contexts. It argues that if recognition and intentionality are basic features of all social interaction anywhere, they also serve as affordances for dealing with, or reflecting on, particular ethical questions that concern a given community. Chapter 4 centers on the problem raised by some of the psychological research, the relationship between those processes that work beyond the scope of the individual's awareness, and what it is people actually think they are doing. It asks what contexts tend to instigate ethical reflections and what resources are available for those reflections. This brings us into a public world of stereotypes, gossip, and other kinds of description, characterization, and evaluation that circulate within a community. It shows how empirical research fleshes out the philosophical idea that people act under the guidance of certain descriptions, frames for making sense of what is going on, what kinds of people are acting, and how actions should be judged. These descriptions circulate in a public world, where part of their power derives from their availability to others. Even if justification, excuses, and blame are post hoc rationalizations of underlying psychological or sociological processes, as the debunkers might say, they come to have a real social existence when they are expressed. They arise in social interactions where these are demanded, accepted, and rejected; here ethics is catalyzed into forms available to other persons. At this point, they are made available for further development, criticism, adoption, or rejection within a larger community. They might become part of a social history of ethics.

The third set of chapters is concerned with the public world and its historical dimensions. Chapter 5 introduces the idea of ethical history. It looks at situations in which hitherto taken-for-granted aspects of everyday life came to be the focus of attention, such as feminist consciousness-raising in the 1960s and 1970s. It argues that processes like this play an important role in the historical transformations of ethical and moral worlds. Chapters 6 and 7 then look at case studies of social movements that took purposeful ethical transformation as one of their goals. Their motives are religious (Christian and Muslim piety movements) in the first of these chapters and atheist (Vietnamese communism) in the second. Juxtaposing religious and atheist movements offers us insight into the different kinds of social conditions that can facilitate, and even demand, the kind of coherence and universality that Williams calls a morality system. One thing piety and communism have in common is the cultivation of a transcendental point of view that can make any apparent ethical inconsistency a problem requiring every effort to overcome. The concluding chapter briefly discusses the problems faced by the universal aspirations of contemporary human rights and humanitarian movements. It then summarizes the case for a multidimensional approach to the study of ethical life in first-, second-, and third-person perspectives, bringing into focus the points of convergence between psychology and the public life of ethical reasoning.

Lurking behind much of the empirical research on ethics, although often unacknowledged, is this question: Why should one be ethical, with what justification? Natural and social scientists may consider this normative question to exceed their brief. Recalling Hume's (1978) distinction between fact and value, for instance, they might reply that their work lies on the side of fact. Yet the question haunts the empirical study of ethical life and often motivates it. For some scientists, the flip side of the causal explanations that seem so debunking may be this: "Because we can't help it—that's the sort of animal humans are." Alternatively, there is the sideways approach of ethnography, to ask how the people we study have themselves answered the question. In the latter case, perhaps the most common answer has been to appeal to ancestral traditions or divine mandates—which, for those who make that appeal, have the status of facts, of a sort. These sorts of authorities were not always as free from challenge as is sometimes imagined, however, and questions about the foundations of ethics have been posed in circumstances as different as ancient Athens and Confucian China (Lloyd 2012). The roots of contemporary scientific research on ethics, however, lie in Europe and America. There, by the end of the eighteenth century, foundational questions could begin to arise, at least in certain schools, salons, and coffeehouses: What, if not biblical command, *are* the sources of and justifications for ethics? A century later, in an intellectual milieu faced with the death of God and the birth of Darwinism, Dostoevsky's Ivan Karamazov could conclude that if there is no God, then everything is permitted, a conclusion whose implications philosophers such as Friedrich Nietzsche would notoriously pursue.

By the twentieth century, both the natural and the historical sciences could be brought to bear on the empirical foundations of ethics. Naturalistic research on ethics and morality often seeks to justify the demands of ethics by demonstrating that it is not contingent but grounded in universal and innate human capacities or by accounting for its origins, thereby showing that ethics is not arbitrary, because it fulfills an adaptive function. These approaches describe a world of causes and effects, mechanisms, or statistical correlations of which people are largely unaware. In contrast, research in history, anthropology, and some related fields tends to seek answers within a world of decision making carried out by self-conscious agents. Although few practitioners in the latter sciences would deny the power of forces that lie beyond ordinary awareness (which might include economics, power relations, demographic effects, and so forth), they usually insist at least on the importance of this: when people act, they have some notion—whether dimly intuited or carefully conceived—of what they are doing. When people act or live in ways taken to be ethical, those notions concern values, exemplary virtues, or ideas about rightness and wrongness. When other people, in turn, respond to those acts or ways of living, they are guided partly by how they do or do not make sense of those ideas. But making sense of ideas is not the end of the story. Ethical life is not just a matter of knowing the rules of the game, something any idle bystander might accomplish as well. It is being committed enough to that game to care how it turns out. A full account of ethical life should help us grasp that too.

CHAPTER 1

Psychologies of Ethics

Seeking Ethical Foundations

One response to the question "Why should one be ethical?" is the turn to naturalism. A version of naturalism would say that being ethical, in the sense of being oriented one way or another to ethical evaluations (which is quite different from being good!), is just part of basic human equipment. This approach may not make ethics obligatory, in the classic philosophical or religious sense, but it seems to grant it the authority of being something fundamental about what people are. The appeal of this position has been summed up by one proponent of "moral realism," the philosopher Peter Railton: "We are natural and social creatures, and I know of nowhere else to look for ethics than in this rich conjunction of facts" (1986: 207). Although naturalism is unlikely to settle the hardest questions, for its proponents it has the virtue of fulfilling the demand that anything we would want to call ethics must be universal and should not be arbitrary. Because some of the strongest claims about human universals have been put forward in psychology, naturalists often turn there for evidence to support moral or ethical realism. Current psychological research on ethics often seeks to demonstrate that it is not contingent but is grounded in innate human capacities. Much of the most well-developed research seeks evidence for the sources of ethics in child development, which will occupy much of what follows, although we will also look at evidence from the moral reasoning and decision making of adults.

Research on the child's development of morality itself has historically been divided between two major approaches. One stresses the learning and internalization of moral concepts and values impressed on the child by the surrounding social world. This has been an especially important approach for scholars interested in cultural differences among morality systems (Shweder 1990). But contemporary researchers on child development tend to reject any strong claim that morality is directly imparted to the child from his or her social surroundings. Instead, much of this work investigates the innate capacities of infants and the child's active role in the emergence of morality. Within this broad area of agreement, child development researchers differ over how much weight to give, respectively, to cognition or to affect, a question that had already divided philosophers such as Kant and Hume. Some developmental models tended to emphasize the emergence of generalized and principled moral concepts out of earlier, egocentric starting points (Kohlberg 1981; Piaget 1965). Others propose that concepts are post hoc rationalizations of more fundamental emotional responses (Haidt 2001).

In the chapters that follow, I will suggest that aspects of both positions are compatible with one another, but neither is complete without putting the individual's ethical capacities into the context of social interaction. But first, let us see where the major findings seem to be heading. Let me be clear, at this point, that most of the research described in this chapter aims at discovering something about human beings in general. Anthropologists and historians will be quick to note that the vast bulk of that research has been with our contemporaries, living in European, American, and a handful of other industrialized and urbanized societies. For purposes of economy, while the reader should bear these limits in mind, I will not qualify every generalization that follows. In subsequent chapters we will take a closer look at some of the historical and cultural complications that general claims about human nature face.

HOW PSYCHOLOGISTS DEFINE ETHICS AND MORALITY

What will count as evidence for the psychological foundations of morality and ethics depends, of course, on how the researcher defines them. Some researchers try to identify a set of components that any moral system will have. For example, some psychological anthropologists have proposed that keeping promises, respect for property, fair allocation, protecting the vulnerable, reciprocity, and the taboo on incest are likely to be universal (Shweder, Mahapatra, and Miller 1990). But there is little agreement among writers at this level of detail. In order to get to a more basic level of understanding, it may be more illuminating to see what the various researchers are defining ethics and morality against. Several pit their findings against the idea that the theory of natural selection must lead us to conclude that humans are driven entirely by self-interested competition (Bloom 2004; Haidt 2001, 2003; Hoffman 2000; Tomasello 1999; Tomasello et al. 2005). None of them reject natural selection. Rather, they propose that the "Machiavellian account" of human evolution must be supplemented one way or another (Tomasello et al. 2005: 688).

From this perspective, the first thing to look for in seeking evidence of ethics in early childhood is unselfish behavior. That means we should be looking for actions that are oriented away from the individual self or that seem to lack any immediate utility.[1] The research tends to accept a distinction between means–ends rationality and those ultimate values that are not means to some further end, a distinction rooted in Western philosophy (e.g., Hume 1975; Kant 1996).[2] Regarding the lack of utility, Michael Tomasello (2009: 64), who has worked with both human children and primates, observes that, although many primates will cooperate to get something that each individual cannot get on its own, only humans enter into cooperative *play*. In this context, play, like art (Bloom 2004), is defined by the absence of gain beyond the satisfaction inherent to the activity itself. It is something that is done for its own sake. It follows that you cannot fully feel that satisfaction unless you are a player or are in some other way committed to the game, like a sports fan. Although this does not mean play itself is ethical, it does show that people can

[1]Neuroscientists interested in morality have found that damage to the frontal cortex sometimes has two effects. A patient who is confronted with an ethical decision will display very limited moral emotions along with an enhanced ability to make utilitarian judgments (Koenigs et al. 2007). One possible interpretation of this finding is that ethics requires an ability to *constrain* purely instrumental rationality.

[2]Some research seems to show that rewarding certain kinds of behavior can make a person less prone to do it. Apparently this "overjustification effect" results when an activity that provides its own reward is devalued by imposing an external reward on it (Tomasello 2009: 9–10).

act without purposes beyond the action itself, which we might consider a precondition for morality.

A perspective on utility and ethics is provided by the well-known "ultimatum game" (Güth, Schmittberger, and Schwarze 1982). In this experiment, one person is given a sum of money and told to split it with someone else. The second person can either accept the offer or reject it, in which case both individuals get nothing. It turns out that if you offer a sum that falls below a certain threshold, around one-third of the total, the recipient will tend to reject it. This flies in the face of the rational calculation of utility, according to which one should accept even a very small sum, since it is still better than nothing at all. One interpretation of this result is that people value fairness for its own sake enough that they would rather forgo monetary gain than be treated unfairly.

One feature of ethics, as commonly understood, is orientation away from the self. For example, Jonathan Haidt defines the specifically *moral* emotions as those "that are linked to the interests of welfare either of society as a whole or at least of persons other than the judge or agent" (2003: 853). Although this is not the only way to define the moral emotions, as we will see, it informs much of the research we are looking at. Even when someone's actions do involve calculations of gain, they count as ethical if that gain is for someone other than the self. In theory, however, one may be immersed in play or in helping someone else without having any particular notion about whether one is doing so for gain or not. This is something an outside observer might be able to decide on objective grounds, with no reference at all to an actor's conscious thoughts or feelings. (Many of the adaptive strategies proposed in evolutionary and game theory have this quality of working behind people's backs.) But most of the psychologists discussed here define full-fledged morality as requiring some degree of awareness. For example, Martin Hoffman writes that "a person's prosocial moral structure is internalized when he or she accepts and feels obligated to abide by it without regard to external sanctions" (2000: 9). This definition, which has aspects of the learning approaches mentioned above, shares with innatist approaches the basic idea that ethics is grounded in an orientation away from the self (one is thus "prosocial" rather than egocentric). It identifies that orientation away from the self with the subjective experience of adhering inwardly to something that one feels is objective and external.

That experience of obligation is often discussed in terms of norms and rules. Like play and art, norms and rules can be treated as values in themselves. According to Tomasello, for instance, children treat even the rules of a game "as supra-individual entities that carry social force independent of . . . instrumental considerations" such as being punished or rewarded (2009: 38). In this view, there are three reasons why a norm might be treated as moral rather than merely conventional. First, its force does not derive from someone perceiving it as instrumental; a rule is a rule even if it does not appear to produce an immediate gain for the person who follows it. Second, the source of the rule lies somewhere beyond the self—while at the same time, one is prompted to adhere to it all on one's own. Finally, one applies it both to oneself and to other people.

So far we have a very rudimentary definition of the domain of the ethical, as that aspect of human activity that cannot be explained either with reference to some further goal or by the force of sanctions. It can be an orientation to other persons or to an activity or a norm treated as a value in its own right, rather than as a means to an end whose value lies beyond the activity. Moreover it has a subjective quality to it: a purely automatic response might not count as ethical (but this becomes complicated, as we will see). Some of these activities are oriented to other persons; others are simply disinterested. All involve some displacement or distance from the self.

There is a second set of features of ethics and morality that is somewhat different. This is a propensity for judgment or discrimination. Research seems to show that from an early age infants sort people out by accent and skin color, favoring some over others (Hirschfeld and Gelman 1997; Kelly et al. 2005; Slater et al. 1998). This discrimination can be a basis for both loyalty and bigotry. Preschoolers, for instance, seem to assume that moral obligations do not apply equally to members of their own allegiance group and outsiders (Rhodes and Chalik 2013). In either case, the impulse to evaluate acts and actors, and to discriminate, seems to work independently of any other function. Other evaluative responses seem to express moral emotions based in reactions of disgust. Like some of the phenomena mentioned above, these responses have most attracted the attention of psychologists when they cannot be explained in terms of immediate utility: to register disgust at rotting food is not likely to be ethical, but to react against an unfamiliar accent is another matter. What Haidt (2001) calls "moral dumbfounding" refers to a person's inability to explain why he or she finds, say, the idea of eating one's pet to be revolting. He takes this to be evidence of underlying moral emotions that are independent of moral reasoning, a point to which we return below. But it is precisely their lack of a clear functional explanation that throws them into the domain of the ethical. Or at least, this is what makes them ethical affordances, that is, candidates for being treated as ethical.

Some philosophers might say that immediate emotional reactions like these lack the full sense of obligation and rightness that defines morality, that they don't give reasons to endorse an ethical stance. Even philosophers who grant an important role to moral emotions, such as Allan Gibbard (1990), limit them to feelings that "it makes sense" to have—anger at injustice, for example, rather than at losing a game. But these responses can develop in ways that afford further moral or ethical elaboration. For example, one set of experiments exposed young children to scenes in which one person either helped or hindered someone else. In one case, infants between six and ten months old watched an animated character either block or assist another's attempt to climb a hill (Hamlin, Wynn, and Bloom 2007). In another, a child who saw an adult grab and tear up another child's drawing looked at the victim with concern (Tomasello 2009: 12). In both instances, the children showed concern for the victim, a clear aversion to the "bad" characters, and a preference for the "good" ones, even when the actions only affected someone else. These responses seem to be the basis for subsequent judgments, among more mature children, about harm or benefit (Haidt 2001: 817; Hauser et al. 2007). To the extent that those judgments concern harm to someone other than the self, it is reasonable to treat them as at least a foundation for ethics. But this does not mean that a full-fledged ethics is where they will necessarily end up. Rather, we can see them as having the potential to be taken up as children mature and develop ideas of fairness, rights, and justice. They also enter into the child's motivation to engage in third-party enforcement, that is, to chastise or punish people who break a rule, even when it has no effect on the self. What both outcomes depend on is some process by which a gut-level response comes to be tied to cognitions that may have quite different sources and functions.

To summarize: underlying much of the research on the developmental psychology of ethics and morality is a basic idea that a core diagnostic of morality is the relative lack of means–ends calculation directed by self-interest. The evidence may include such things as empathic responses to another person's pain, or cooperation, or play. But recall that Shweder's list of possibly universal ethical principles includes a prohibition on incest. How does this fit in? To take this into account, we need to include the second diagnostic, judgment, a basic propensity to evaluate people and their actions as good or bad. Conjoined with the orientation toward social norms (thus, away from the self), the propensity to evaluate can be taken up by particular practices of child socialization to local cultural values. Norms, such as those of kinship, have a social history that involves cognitive processes of quite a different sort from the neurophysiology of the moral emotions. But those emotions can be recruited in ways that give more punch to what might otherwise be a fairly dry set of rules.

The revulsion against incest is an example of Haidt's moral emotions, which might be innate but which take their particular content by being linked to a specific set of learned norms.[3] In short, two basic orientations to others emerge out of the psychological research. One is empathy and cooperation; the other is evaluation and judgment. What these have in common is that the child experiences them as not coming from him- or herself.[4]

[3]One of anthropology's most well-established findings is that the lines dividing kin from non-kin vary widely across societies. For a man to marry his maternal first cousin is incest in some kinship systems and a highly preferred alliance in others, such as the Sumbanese mentioned in chapter 1. From this, many kinship specialists conclude that even though the apparent *universality* of the incest prohibition might be explicable purely in terms of some social or biological function, its distinctly *ethical* weight and scope depend on particular social histories (Lévi-Strauss 1969).

[4]One might ask whether a very young child really has a sense of self at all, but the definitions of morality presumed by the research discussed here seem to depend on that being the

EMPATHY AND ALTRUISM

Does this displacement away from the self mean that children are innately disposed to altruism? Here we need to be careful not to immediately conflate all forms of empathy and cooperation with the morality of altruism. As noted above, cooperation for mutual benefit might just be an intelligent form of self-interest, to which benefiting others is secondary. Consider empathy. At a very early age, infants respond to other babies' signs of pain (Bloom 2004: 119). This might not yet count as empathy. Hoffman defines empathy as "an affective response more appropriate to another's situation than one's own" (which might include joy as well as suffering) (2000: 4). This response depends on being able to differentiate your own feelings from another's: if I automatically feel pain when I see someone else hurt, I may just be responding to my own pain. According to Hoffman, empathic arousal in early childhood is, in effect, the involuntary experiencing of another's distress without realizing that it is not one's own (something like what Pfaff seems to have in mind in his neuroscientific account of altruism, mentioned in the introduction: a momentary loss of distinction between self and others).

Empathy can be defined by a three-way distinction among knowing that another is in pain but not experiencing suffering oneself, suffering because of another's pain but not distinguishing between that person's pain and one's own, and suffering on behalf of another (the same, of course, applies to empathic responses to another's joy). Researchers seem most confident in linking the last of these to ethics (Bloom 2004: 113; see Hollan and Throop 2011). A more developed empathic response has been mentioned above: a child who sees an adult destroy another child's drawing will look at the victim with concern. Here we see evidence of attention to another person that suggests the distinction is meaningful.

---

case, at least by the point in development in question in a given case. I thank Susan Gelman for raising this question.

The developmental psychologists who see empathy as an important foundation for the development of morality note that it is not sufficient. In a morality system, at least, you can act morally without empathy (giving to charity on principle) and may have to suppress empathy in order to act morally (the police officer who knows that she makes the criminal unhappy but arrests him anyway). Empathy does not necessarily lead to compassion. Empathically strong ties to one's own group may reinforce aggression against outsiders (Hoffman 2000: 21). Paul Bloom notes the case of a woman in Nazi Germany who lived near a death camp. The sounds of the executions disturbed her (an empathic response), so she petitioned the camp's commander to carry them out somewhere else (hardly a strong case of compassion) (Bloom 2004: 117).

Now let's look at helping. Tomasello (2009: 14) observes that great apes do not point in order to inform other apes of something. They point as a kind of imperative, a demand to be given something for themselves, which they do only when they stand to gain something immediately. Only humans point simply in order to impart information or otherwise share a perspective with another person without the prospect of some immediate personal gain, other than, perhaps, sociability itself. The urge to be sociable for its own sake appears very early in infants. By the age of eighteen months, toddlers are spontaneously helping others (Warneken and Tomasello 2006). It seems that this propensity to help gets taken up as people develop more purposeful kinds of sharing and cooperation—not because in itself the impulse to help necessarily results in sharing.

Is this altruism? Most of the researchers would say that an automatic response to certain kinds of stimuli, such as nursing a newborn, is not yet altruism. The usual meanings of English-language uses of *altruism* and *morality* seem to require some role for subjective awareness and motivation (Kitcher 2011). Following Tomasello (2009: 57), these depend on the prior development of the cognitive capacity to take another person's point of view. What seem to be crucial to all these developments are some fundamental cognitive and emotional capacities to be aware of your pain but also know it's not mine. If empathy is one precondition for ethical life, it depends on the development of certain ways of both identifying the self with and distinguishing it from other persons.

SELF AND OTHER

We have seen that cooperative play is an activity carried out for its own sake. It is also evidence of an ability to grasp and manipulate the identity of and difference between self and other. There is no clear consensus on the developmental history of selves in relation to others. One tradition, which goes back to the psychologist L. S. Vygotsky (1978), takes the infant's awareness of being distinct from other people to be something that is not given but requires a process of differentiation over time. An alternative approach looks at a reverse process, by which the self learns about others' subjective experiences by drawing on its own (Meltzoff and Brooks 2001). As I will suggest, it may well be that each perspective is right in certain ways.

Most of the researchers discussed here agree with some version of the innatist view that infants are endowed with certain propensities and motivations that lead them to seek out interactions with others virtually from birth. These interactions in turn are critical to the development of further capacities. Some research shows that newborns can distinguish between animate and inanimate entities and are predisposed to attend more to the animate ones—things that will respond to them. As his or her physical abilities mature, the young infant may try to instigate responses from others through gaze, sound, and gesture, but further neurological development depends on what those other people do. By the age of two months, infants are already engaged in patterns of interaction with caregivers that allow them to enter into the rhythms of adult conversations (Trevarthen and Aiken 2001).

Newborns imitate the bodily movements of other people in ways that seem to indicate that they are able to "map equivalences between self and other" (Meltzoff and Brooks 2001: 173) as well as between the subjective states of the self's own body and the visual perception of those of other people. Responding to and in turn prompting these imitative impulses, people around the world play mimicry games with babies. According to Meltzoff and Brooks, these games promote an ability to identify what one sees others doing with what one feels and does oneself. These games depend on the infant's response to others' intentionality: "Only people who are paying attention to you and acting intentionally can match the form of your acts in a generative fashion. Only people can act 'like me'" (Meltzoff and Brooks 2001: 178).

One implication of these findings is that any innate propensities the child may possess cannot in themselves determine their eventual outcome. Even their neurological development may require uptake by other people. That requirement makes the ongoing engagement of other people not just something the self has to deal with, a matter of social context from which the self can be extracted. It is an essential element of what makes up a self at all. As we will see in the next several chapters, this is crucial because ethics is not just something I do to, or with regard to, other people who stand outside me. It is a function of their constitutive presence within my sense of self and mine in theirs. I mean this not in some mystical way but as the outcome of developmental processes and the kinds of social interactions that persist through adulthood.

Toddlers around nine to fourteen months old go beyond bodily movement to engage others in joint attention. They can follow another person's gaze and pointing gestures, interpreting them as evidence of something mental. In cognitive terms, this requires two things. First, it means that the child can extrapolate from another person's gestures to something that is not directly visible, such as attention or thought. Second, in order to focus on what is in the other person's visual field instead of his or her own, the child must be able to withdraw his or her attention away from the immediate contents of his or her own visual field—that which is most real to his or her self—and shift focus to what is central to someone else's. This decentering skill is not as obvious as it might seem: being unable to do it is one symptom of autism (Baron-Cohen, Baldwin, and Crowson 1997).

But the reverse may also be the case. For example, R. Peter Hobson, a specialist in autism, is skeptical that children can project from their own mental experiences directly onto the expressions and behavior of other people. In his view,

> far from reflective self-awareness preceding the ascription of mental states, it is rather that some form of awareness of other minds is a precondition for acquiring reflective self-awareness. The reason is that in order to adopt a psychological orientation towards one's own mental states, one needs to appreciate that there *are* alternative vantage points that one can assume towards a mind-endowed self, and for this the experience of psychological "co-orientation" with others may be necessary. (1991: 46; see also Hobson 2004; Tomasello et al. 2005: 689)

This is not the place to settle the debates among the psychologists, but one reasonable hypothesis would be that both directions of projection and inference work, through a kind of looping effect at different points over the full course of a child's development.

Certainly it seems that being able to engage in shared intentionality even at a rudimentary level depends on each individual's ability to swap roles with others, enter into their point of view, and have goals in respect to their goals (Bratman 1992). The capacity for role reversal has to be in place by the age of thirteen or fourteen months. It is a precondition for language learning, which depends on what linguists call "shifters" such as the pronouns *I* and *you*, which change whom they refer to with each turn of talk (Benveniste 1971). This fosters the more general ability to take multiple points of view:

> To comprehend the linguistic communications of others, children must in some sense simulate the perspective of other persons as they are expressing themselves linguistically, and so the back-and-forth of discourse involves the child in a constant shifting of perspectives from her own to those of others and back again. (Tomasello 1999: 176)

The reciprocity of perspectives is also a precondition for other basic conversational abilities such as turn-taking, sharing perspectives, and the ability to clarify points of misunderstanding. But these interactive skills may loop back. Internalizing the voices of others may provide resources for reflection in the form of inner dialogue with oneself (Haidt 2001; Tomasello 1999; Wertsch 1991). At the same time that the self is learning to internalize its interactions with other persons, as Adam Smith (1976) suggested long ago, it is learning to see itself from the outsider perspective of those persons as well. Tomasello et al. hypothesize that

> in understanding an adult's intentional actions . . . at the same time that she experiences her own psychological states toward the other, the child comes to conceptualize the interaction simultaneously from both a first and third person perspective—forming a "bird's-eye view" of the collaboration in which everything is comprehended in a single representational format. (2005: 689)

And this distancing or outside perspective facilitates not just concepts but also judgments. Internalized voices may be an important source of moral authority in inner dialogue (Tomasello 1999: 194; Tomasello et al. 2005: 684). This is why even apparently private ethical thought draws on conversations with other people, a point we will return to in later chapters. But the reciprocity of perspectives is not only a resource on which ethical and moral reflection can draw. Tomasello also suggests that the initial identification of self with other on which this reciprocity builds itself is facilitated by "the skills and motivations for interpersonal and emotional dyadic sharing characteristic of human infants and their caregivers" (Tomasello et al. 2005: 689). In other words, through a kind of looping process, early emotional bonds between adult and child may prompt forms of self–other identification that in turn can become available for the development of greater ethical self-awareness. Of course this development has a dark side as well since, as noted above, ethical development typically also entails learning discrimination, knowing what kinds of persons one does *not* identify with, even those to whom moral norms do not apply. As we will see below, the ability to shift among first-, second-, and third-person perspectives is an important element of ethical life in both face-to-face interaction and the dynamics of social movements.

MIND READING

You can only aim to be helpful to someone if you are aware of his or her goals as distinct from your own. Helping therefore depends on aspects of the self/other distinction that start to appear early in child development. Young infants make a distinction between entities that seem to have agency—and thus might be relevant for interaction—and those that do not. By nine months, a baby can tell the difference between intentional and unintentional actions and, by fifteen months, can identify the goals of someone else's actions (Csibra 2003; Meltzoff and Brooks 2001; Onishi and Baillargeon 2005; Stone 2006; Tomasello et al. 2005). These are preconditions for helping someone whose attempts to achieve a goal have visibly failed, for example, losing a toy behind the couch or throwing a ball at a basket and missing (Bradone and Wellman 2009). This distinction between intended and unintended actions, which marks that between goals and actual behavior, also underlies the way children who are watching other people interact with each other end up evaluating them. The child has to know one person's goals in order to judge whether a second person is helping or hindering—and therefore whether that other person is good or bad (Hamlin, Wynn, and Bloom 2007). But for the kind of awareness that gives helping its fully ethical character, the child must also be able to tell the difference between another person's goals and his or her own, something that becomes possible at around eighteen months (Repacholi and Gopnik 1997).

Therefore, at the heart of the helping relationship is some capacity for what we can call "mind reading," the child's developing ability to impute to other people feelings, thoughts, and knowledge that are different from his or her own feelings, thoughts, or knowledge. One of the things that mind reading allows is the shared intentionality that becomes possible at around fourteen months. This refers to a distinctly human capacity to collaborate by people who are so closely coordinated in their goals and plans, but also responsive to one another, that they can be properly called "we" (Tomasello et al. 2005). Between the ages of three and four years, a child has generally developed the ability to infer (accurately or not) others' intentions, goals, desires, and feelings and finally their beliefs about states of affairs, such as whether there are cookies hidden inside a container (Wellman 1992). All of this also requires a kind of self-distancing, the ability to suppress what is in one's own mind or perceptual field in favor of what is in someone else's (Cacioppo, Visser, and Pickett 2006).

What are the implications of these findings for our understanding of ethics? There are two things to observe in this context. One is that mind reading is an important precondition for ethics in more than one way. It makes possible helping and judging others' helpfulness. It also lies behind all sorts of ways people hold others responsible for their actions. For example, children who do not yet have full mind-reading abilities usually lay equal blame on someone who throws out their cupcake on purpose and someone who does so by accident; those who do read minds normally find the latter less culpable (Killen et al. 2011). Similarly, mind reading underlies the distinction between saying something in error and lying (Sweetser 1987).

On this basis, more complex variations on responsibility become possible. For instance, children at a certain degree of maturity begin to distinguish between mistakes of fact and mistakes of law (Chandler, Sokol, and Hallett 2001). This distinction holds that you are less culpable for breaking a rule if you were not aware of all the facts than if you did not apply the rule properly. Underlying this notion is the idea that you have a responsibility not just to follow the rules but also to try to interpret them correctly. Now this distinction relies on a specific legal tradition and may not apply across all societies. The point here is simply that as a person's skills at mind reading grow, the range of intentionality that he or she might take into account does as well, which can lead to more nuanced ethical judgments.

But none of this means that mind reading necessarily produces ethics. The judgments mentioned here draw on a disparate set of affective and cognitive processes, involving distinct parts of the brain and serving a wide range of functions, only some of which we might want to call ethical (Cacioppo, Visser, and Pickett 2006; Greene and Haidt 2002; Greene et al. 2001). Put another way, we cannot look directly at a specific capacity of mind in order to identify a basis for ethics. Ethics draws on those capacities, but what makes something count as ethical and not, say, as a means to a certain practical end (such as knowing how to cooperate with someone in climbing over a fence) must have some other sources. We will see in the chapters that follow that those sources can be found in processes by which people in their interactions with one another take up these basic capacities in light of ethical projects or problems. But before we take that step, we need to fill in some of the other aspects of the ethical capacities and some serious challenges that psychological research poses to the very idea of ethics.

PSYCHOLOGY'S CHALLENGE TO ETHICAL AWARENESS

As noted above, some research finds that from a very early age, children will readily evaluate other people on the basis of whether they are assisting or hindering someone other than themselves. But judgments about other people are not always based on questions of help or harm. For instance, we have also seen that infants show preferences for people in categorical terms that seem to sort them into those more or less like themselves in some respect. In adulthood, people continue to judge and categorize people instantly and automatically (Haidt 2001: 819–20). Like cooperation and mind reading, evaluations such as these are not in themselves necessarily ethical, but they can afford responses that are.

The very quick judgments and evaluations that appear early in childhood are primarily emotional in nature. That is, they do not result from the child consciously thinking things over and making decisions. In this respect, they seem to be one manifestation of a broader phenomenon. According to Haidt (2001, 2003), emotion is a better predictor of people's first ethical responses to a situation or person than judgments of harm. Not only does moral reasoning, he says, show up later in a person's development (which is hardly surprising), but even at that point it still plays a secondary role in the ethical life of mature adults. What is most fundamental, in his view, is what he calls the moral emotions.

In a famous thought experiment, Haidt's team presented people with a series of scenarios. These included an act of brother–sister incest, eating one's dead pet dog, cleaning one's toilet with the national flag, and eating a chicken carcass one had just used for masturbation. The scenarios were presented in such a way that it would be clear that no harm would result from any of these actions (for instance, the incest was consensual and nonreproductive). Nonetheless, people's usual reaction to each scenario was quick and unambiguous: the act is simply wrong. But when asked to explain why it is wrong, people often had trouble answering, saying things like "I don't know, I can't explain it, I just know it's wrong" (Haidt 2001: 814). Haidt calls this gap between the ease of the response and the difficulty of rationalizing it "moral dumbfounding." He takes it as evidence for the existence of fundamental moral emotions, which are fast and automatic, in contrast to moral reasoning, which is slower and requires effort. Reasoning in cases like these is an after-the-fact and somewhat ad hoc attempt at justification for a response that has already occurred.

Haidt gives a neurophysiological version of this developmental story based on relations between distinct components of the mind. In broad outline, it resembles an ancient model. Plato's *Phaedrus* 246a–254e (1961: 493–500) portrays reason as a charioteer trying to control two steeds, one of which is the unruly passions. David Hume also placed passion and reason in opposition to each other, although he (1975: 415) reversed their proper roles, holding that reasons should be slave to the passions. And Freud's (1949) topology draws a distinction between the controlling superego and unruly id. Consistent with current efforts to identify different psychological phenomena with specific brain functions, Haidt suggests that the fast and automatic "hot" responses he calls moral emotions involve the amygdala, a region of the brain associated with (among other things) fear and other strong emotions. Only later, as a child matures, do the hippocampus and frontal lobe introduce the inhibitions and executive control that are needed by "cool" reasoning. As evidence for this distinction, Greene and Haidt (2002: 518) point to cases of adults with brain damage or psychopathy who clearly understand moral reasoning but whose highly immoral actions seem to be unaffected by it.

As in the older models, the distinction between hot and cool also marks one between degrees of self-awareness. In Haidt's models, the workings of moral emotions, unlike reasoning and justification, are largely unavailable for self-inspection. Their effects may be registered in one's awareness but not their sources. In this respect, the distinction between hot and cool is part of a more general challenge that psychological research poses to some familiar ideas about ethics. Recall the common position, noted in the introduction, that ethical actions are defined, in part, by the actor *knowing* that the act is ethical. Much of the psychological research we have been looking at in this chapter involves processes that either work beyond the actor's self-consciousness or, like moral emotions, enter awareness without any particular ethical justification. Where stronger versions of self-awareness do enter the picture, they are often explained with reference to some underlying processes in which awareness plays no part.

In a book devoted to this problem, the philosopher Anthony Appiah sorts the psychological findings into two fundamental challenges to moral philosophy. One is posed to the reality of personal character; the other, to the validity of ethical intuitions. The first challenge centers on an important assumption in virtue ethics, as well as much contemporary Western folk psychology, that an individual has a stable character, with certain dominant virtues and vices, which is consistent and therefore more or less reliably predicts his or her behavior. This view of character has been challenged by so-called situationism, the claim that a person's ethical choices are determined by factors that are beneath the level of awareness and do not in themselves involve ethical values (Ross and Nisbett 1991). Rather, "a lot of what people do is best explained not by traits of character but by systematic human tendencies to respond to features of their situations" (Appiah 2008: 39). Small differences of context can lead you to quite opposite intuitive responses.

Appiah (2008: 41) gives the example of one characteristic finding, that people are more likely to get a stranger to help them change a dollar if they are standing near a bakery than in a less fragrant spot. Another study (appearing after Appiah's book) reports that people asked to think about a scenario with their eyes closed judge immoral behaviors as worse and moral behaviors as better than those doing so with their eyes open (Caruso and Gino 2011). In general, situationism holds that apparently trivial or ethically irrelevant differences in the immediate situation can have remarkably strong consequences for people's choices. In other words, people turn out to be far less consistent than one would expect if their actions were governed by their character. We will return to the problem of character in the next chapter. Here consider the second challenge.

Like situationism, what Appiah calls the "challenge to intuition" points to evidence of cause-and-effect relations that work beyond the individual's awareness. People's reasoning may be biased by the way a moral problem is posed. For example, which sequence you present the options in invokes "order effects," since earlier options tend to be favored over later ones (Appiah 2008: 83–92). Similarly, "prospect theory" refers to the bias introduced by whether one frames identical situations by stressing the possible losses or the possible gains that might result from a given course of action (Kahneman and Tversky 1979). Differences in phrasing that seem to have nothing to do with ethics as such can lead to diametrically opposed responses to objectively identical situations. Although situationism includes affective responses, whereas the challenge to intuition centers on cognitive processing, both sets of findings have certain things in common, notably that consequential differences in how people act seem to be due to causes that in themselves are both "small—and morally irrelevant" (Appiah 2008: 40).

This characterization depends on the analyst knowing what counts as small or large and what counts as morally relevant or not. As we will see, when we turn to the distinction between ethics and social conventions, matters of weight and relevance may not always be clear-cut. Nonetheless, these findings are significant because they seem to undermine the actors' own sense of what is trivial or not and what is ethical or not. In this light, the most important challenge they both pose is to the ideas that people know what they are doing, that ethics depends on that self-knowledge, and that this self-knowledge should affect how people act. In short, they undermine the idea that ethical reasoning makes a difference in what people do. Indeed, this kind of research may raise general questions about the role of people's self-awareness in governing their actions. The challenge, then, involves two claims: that self-awareness has no causal effects on one's actions and that the real causes of behavior elude consciousness. In this respect, as I suggested in the introduction, these claims are of a species with the more general challenge that much scientific research has posed to ethics and morality, by introducing forms of causality that seem to work independently of, and possibly at cross-purposes to, the realm of values and reasons.

These are the challenges that Appiah and the psychologists all agree on. Here, however, I want to point out another aspect of these findings. Appiah (2008: 96) remarks of the trolley problem, described in the introduction, that it is a moral emergency. His point can be applied more broadly. Not all the scenarios used in situationist experiments are emergencies, but they do have several features in common with them. They usually involve individuals, acting on their own, making either/or decisions, with clear options, within a limited period of time, with unambiguous outcomes, and with no aftereffects, such as subsequent remorse. This characterizes a great deal of the experimental research on child development as well. We may well wonder whether these features really capture what most of ethical life is like. Certainly Bernard Williams and others who approach ethics as a lifelong project, rather than a matter of momentary decisions of right and wrong, would say no. Since people *are* self-aware, even if we grant that these findings expose the limits to that awareness, we might still ask what role self-awareness *does* play, if we agree that it's not that of sovereign self-governance. Since ethical self-awareness often refers to norms and values, we can start by asking what the relationship between emotions and norms is.

MORAL EMOTIONS AND NORMATIVE JUDGMENTS

In Haidt's depiction of the moral emotions, rationality plays a role that is merely secondary to the moral emotions, offering after-the-fact justifications and rationalizations for judgments that the person has already made. Interestingly, Hoffman, working with observations of child development, draws what at first looks like the opposite conclusion. Generalizing from his findings, he says that moral principles are usually first conveyed to children in "'cool' didactic contexts" and only later, as those children become committed to them, do these principles "become emotionally charged . . . 'hot' cognitions" (2000: 15). One might imagine, for example, a child who learned the meaning of the national flag amid the tedium of the school day becoming the adult who is prompted to visceral rage by the idea of flag-burning. But this might in fact be consistent with Haidt's proposal. The didacticism to which Hoffman refers would normally come into play only after the child's cognitive functions are sufficiently developed to handle "cool" reasoning. Once they come to possess the necessary executive control, children are able to learn from explicit, rationalized talk about morality, the stuff of parental admonishments, legal codes, the precepts of the elders, or the dictates of religious training. But for these to be truly effective in shaping the child's ethical life in practice and not just theory, they must be connected back to the more basic moral emotions. Those emotions provide motives and commitments that mere principles might be unable to generate. Conversely, those principles provide the emotions with objects that one can endorse—they help explain why it makes ethical sense to have those emotions (see Gibbard 1990). If this interpretation is right, then that looping effect is evidence of an ethical affordance, in which a social project of establishing moral reason takes advantage of the emotional propensities that are already in place. However that may be, the relations between hot and cool that Haidt and Hoffman describe serve an indefinite range of possible functions; nothing determines that they will necessarily serve ethical ones.

The sort of didacticism to which Hoffman refers teaches social norms and usually does so quite explicitly. To understand the distinction between norms and the ethical intuitions, consider this: repugnance at eating one's pet dog might simply be a matter of preferences, like disliking linoleum floors or classical symphonies. Liking or disliking something hardly counts as ethics or morality as such. By that standard, simple matters of taste would also be ethical ones. Why should we consider the moral emotions to be *moral*? True, they have one feature of ethics as the psychologists have been using the concept, the sense of a fundamental value judgment that is not a means to some further end (note, for instance, that when Haidt asked his subjects whether an act of incest was immoral, he made sure that the scenario specified that birth control was used, eliminating any objections on the grounds of genetic harm). More, however, is needed. In the case of the social norms that are taught didactically, what this consists of is reference to values or categories that are recognized by other people: "*Moral judgments* are therefore defined as evaluations (good vs. bad) of the actions or character of a person that are made with respect to a set of virtues held to be obligatory by a culture or subculture" (Haidt 2001: 817).

We will return to the question of culture in later chapters. The point to bear in mind here is the role played by other people's ethical judgments in stabilizing or rationalizing one's own. This too draws on two basic human propensities. One is conformity; another is the proactive motivation to seek out norms. The existence of social norms depends on people's tendency to engage in "imitation for purely social motivations—not just to accomplish goals but to be like others" (Tomasello et al. 2005: 687–88). More than this, by the age of three children also seem to be drawn toward seeking out norms. When entering a new situation or game they are prone to *assume* that there are customs or rules to follow, and they take the initiative in figuring out what they are (Kalish 2006; Rakoczy, Warneken, and Tomasello 2008). They do so even when there is no authority, such as a parent or teacher, to dictate or enforce the norms.

Norm-seeking is especially interesting in light of the philosophical distinction, famously articulated by Hume (1978), that there are no logically valid deductions from what "is" to what "ought" to be. The child's norm-seeking behavior seems to run against this dictum, since it is a way of eliciting oughts (the rules of the classroom or the game) from the observation of what is (how people act in the classroom or when playing a game). The child is not just interested in how other children *are* playing a game or entering a classroom; he or she wants to know how they *should* be doing so. But the only way children can grasp what "ought" to be is from their perceptions of what "is" the case. There is a distinctive kind of inference involved. The child must extrapolate a rule, which itself is not visible, from what other people are doing and sort out which behaviors manifest the rule and which are irrelevant. In doing so, according to some philosophers (e.g., Wittgenstein 1953), children are going beyond what is logically justified by the empirical evidence. From the perspective of psychological research, this very excess seems to offer evidence that the child's assumption that there must be rules or norms to be discovered is innate.

Although norm-seeking might be a manifestation of an innate impulse toward conformity, it is not identical to it. For one thing, it seems to be detachable from the self's own behavior or need for approval. Once a child has acquired a sense of what the relevant norms are, in any given situation, he or she is also prone to engage in what is called "third-party norm enforcement." For example, if one player tries to alter the way a game is being played, three-year-olds will object strenuously on principled grounds: not just that the alteration is different from what they had been doing but that it is wrong—that is not how the game ought to be played. More generally, in Tomasello's (2009: 39–41) view, when children enforce norms on other children (for instance, admonishing one child not to take another's dessert), this shows not only that they are trying to conform for their own benefit (perhaps in order to fit it) but that they value the norm as such. Here again it seems that displacement from the self is central to some basic understandings of ethical life.

At this point one might object that I have conflated moral rules and social conventions. After all, the rules of a game and classroom behavior hardly have the same status as the biblical injunction against murder, Kant's Categorical Imperative, or *Star Trek*'s Prime Directive. Some researchers claim that children universally are able to distinguish between morality and convention. For example, the developmental psychologist Elliot Turiel (1983) defines morality as obligations that concern objective harms and the welfare of others and convention as right and wrong actions defined arbitrarily by social consensus. He holds that children in every society recognize the distinction between them. For instance, children will acknowledge that it is wrong to hurt another person, regardless of any external authority, but that the classroom rule about raising your hand in order to be called on by the teachers could be changed and therefore does not have the same status.

Ethnographic research, however, shows that the universalizing claim is hard to sustain, for at least two reasons. One has to do with differences in ontology, what people take to be real. What might look like mere social convention for one person—for example, wearing a beard and tefillin and abstaining from pork—is divinely sanctioned scriptural mandate for another—in this case, an Orthodox Jewish male. The psychological anthropologist Richard Shweder and his colleagues argue that no substantive version of the morality/convention distinction holds across cases. For instance, "among orthodox Brahmans and Untouchables in India, eating, clothing, and naming practices, and various ritual events are viewed in moral, rather than conventional, terms" (Shweder, Mahapatra, and Miller 1990: 159). The authors conclude that it is unlikely "that there exists a universal class of inherently nonmoral events. . . . Any event can be made moral by appropriately linking it to a deep moral principle" (1990: 160).

Where distinctions are made, their boundaries can be very fuzzy. Many American Mormons, for example, have a sense of dress appropriate to the pious that does not quite reach the level of mandate but still exceeds that of mere convention (Kramer 2014: 27–31). More generally, it seems that as children come to understand people in terms of social categories, they come to treat those categories as normative, involving rights, duties, and expectations; knowing that someone is a doctor or teacher leads to normative assumptions about how that individual ought to behave (Hirschfeld 2013; Kalish and Lawson 2008).

Even in societies where a distinction between morality and convention is recognized, adherence to convention can still indicate a person's moral fiber. Practices that people acknowledge are conventional in themselves, such as ways of dressing, forms of greeting, or terms of address, might be linked to more fundamental moral principles. The linguistic anthropologist Steven Caton reports that in Yemen, where men must constantly juggle the demands of honor and piety, sometimes in tense or even dangerous circumstances, the act of greeting one another carries enormous ethical weight, without which a social interaction simply cannot get off the ground. Common formulas for greeting, such as "May your morning be blessed," make clear that "when greeting a person one is in a very real sense engaging in a religious act, calling on God to bestow his favor on the addressee" (Caton 1986: 294). In the details and variations in how he carries out this simple act, the speaker embodies his own piety and, by implication, honors his addressee by granting him a similar standing—or dishonors him by muting it. We will go into the nature of this kind of self–other dynamic, and its expression in deference and demeanor, in more detail in the second section of this book.

In many societies, even if there is no rule mandating proper dress, food, or speech, being willing to breach what are acknowledged as merely customary norms may nonetheless be taken to say something about one's moral character. Making the point even stronger, Chandler, Sokol, and Hallett assert, regarding truth and moral rightness, that "all of us, along with the institutions and social practices we have created in order to manage our affairs, regularly proceed as though these two domains were deeply connected" (2001: 349). In saying this, these psychologists are pointing to a crucial point of articulation between those impulses and capacities that seem to be innate to young children everywhere and the particular social worlds they are entering, ones whose concepts of reality and moral order may vary widely. But we must take things one step at a time. First we need to look more closely at the cognitive and affective emergence of the self-distancing and displacement characteristic of ethical life.

THIRD-PERSON PERSPECTIVE

Several features mark the propensities and capacities evident in child development as having a distinctively ethical potential. These include a strong orientation toward other persons for their own sake, an ability to displace one's attention and feelings away from the self, helping and cooperating behavior, gut feelings of disgust or attraction, the motivation to engage in activity without instrumental goals beyond the activity itself, the propensity to evaluate persons and acts on other-than-utilitarian grounds, conformism, and norm-seeking. This is quite a diverse collection. The propensity for evaluation, for example, may overlap with motivations that are independent of self-interest but does not have any obvious connection to helping or cooperating with other persons. As I will suggest in the chapters to come, if ethics is a unified category, this unity does not derive from a single shared property of the child's basic capacities and propensities. Rather, the coherence of "ethics" in any given instance is constructed through social processes. Social interactions and the projects that enter into them take up the ethical affordances that appear in childhood and make them into something that comes to be recognizable to the participants as ethical, given a particular social context.

But the orientation to norms, which can draw on the moral emotions and the evaluations they give rise to, does eventually converge with the developing relations of self and other in certain ways. Consider some of the roles that other people play in the self's development. Other people serve children as models for imitation, as the recipients of helping actions, and as the objects of stance or evaluation from early in life. Since cooperation requires that one be able to orient one's own goals with the other's goals, it depends on an ability to swap role perspectives. Similarly, language acquisition depends on being able to swap speaking roles, for instance, through turn-taking and the alternation of pronouns. Third-party norm enforcement marks one's orientation to a community at large, rather than just the particular individuals with whom one is interacting at the moment. We might conceptualize this as the expansion from first person (I/me) and second person (you) to first-person plural (we) and on to third person (he, she, it, they). In English, the third person is marked by its position outside the immediate interaction of speaker and addressee. It stands for an open-ended field in which the status of anyone at all might be located.[5]

Consider a child who chastises another child for taking a dessert from a third one. According to Moll and Tomasello (2007), third-party norm enforcement like this exemplifies shared intentionality. That is, the commitment to norms even in the absence of self-interest (it wasn't the child's own dessert that was taken) is due to the child's commitment to the collective "we" whose shared intentions the norm represents. Shared intentionality ultimately requires the cognitive ability to grasp not just the expanded first person (we) but also the third-person perspective (Kalish 2012). This is because the child must be able to track both his or her own actions and purposes and those of another, to remain aware of the differences between them, and to grasp how they fit together. This latter ability is the outsider's view, which can be called the "third-person" perspective. It is not just taking another person's point of view, which would simply replace one first-person perspective with another. Fully developed, this means being able to take a generic view, a perspective from outside the immediate situation. Norms, in the generic view, apply

[5]Pronoun systems vary widely among languages, and the English version here is only meant as a heuristic. However, it is the case that all languages have pronoun systems that must minimally distinguish among speaker, addressee, and people outside the conversation (Silverstein 2003a).

regardless of who is or is not playing the game or participating in the classroom. Similarly, social roles such as doctor or teacher are linked to expected behavior, regardless of who inhabits that role.

Eventually, the ability to take a third-person perspective facilitates the person's skills at reflecting on that very self that has been the subject of the first-person perspective, as "self-observation employs all of the categorization and analytic skills that are employed in perceiving, understanding, and categorizing the outside world" (Tomasello 1999: 195). The full flourishing of this ability seems to depend on the mediation of language. Through what Tomasello (following Karmiloff-Smith 1992) calls "representational redescription," the first person's otherwise tacit grasp of things, from immersion in the midst of the action, is externalized and made explicit in ways that allow the thinker or speaker to see things the way some other person, such as a "third person" external to the situation altogether, might see them as well. As a cognitive capacity, the third-person perspective fosters greater flexibility in applying concepts and categories. Such concepts and categories typically involve some degree of abstraction and generalization of what might otherwise be very specific, concrete circumstances, the results of which can be applied more widely.

They might also interact with other processes. There is some evidence that people's responses to harm, for example, involve different regions of the brain, depending on whether the violations feel close-up and personal or relatively impersonal. This suggests, at least to some researchers, that whereas personal violations may prompt automatic domain-specific reactions, impersonal ones give rise to more general reasoning processes (Greene and Haidt 2002). This may help explain why, in the trolley problem, people are more unwilling to push the fat man off the bridge, as in one scenario, than to divert the trolley so that it will hit the fat man on the tracks, as in the other. In either case, their action results in the death of a man. Their differing reactions seem to involve physical proximity (people are more resistant to pushing someone in person than pulling a switch). But their gut feelings may also endow more conceptual distinctions, such as that between killing and letting die (one aspect of the Thomist principle of "double effect," that some harms are permissible as the unintended consequences of acts aiming at a greater good), with some of their emotional impact and the sense that they are intuitively right (Foot 1967).

The ability to think in terms of generalizations and abstractions is not just a matter of individual cogitations. As suggested, in its fully developed state it seems to require language. The specific linguistic forms that facilitate the third-person perspective, while dependent on universal cognitive capacities, have been produced over historical time through the interactions among large numbers of people. In the chapters that follow, I will suggest that this history and those other people must therefore, in some way, be considered necessary conditions for the individual's ability to take up a third-person perspective, to generalize, and even to see his or her own actions from an external point of view. What "in some way" means, of course, is a key question, something to which we will be returning from various directions over the rest of this book. Description makes possible such things as appealing to general principles (life, liberty, and the pursuit of happiness), entertaining counterfactuals (what if there were no rules against murder?), and attending to things beyond the perceptually immediate place (I am concerned about starvation in Sudan) and time (history arcs toward justice). As Hoffman puts it,

> The ability to represent thus expands the importance of empathetic morality beyond the face-to-face encounters of children and members of primary groups.... It expands the bystander model to encompass a variety of situations limited not by the victim's presence but by the observers' imagination. (2000: 8)

The expansion of reference and the taking up of a third-person perspective that descriptions make possible are sometimes linked to certain developmental narratives (themselves echoing themes of Enlightenment philosophers such as Adam Smith) in which the maturing individual gradually expands the circle of his or her moral concern. In Hoffman's model, for example, children start out feeling egocentric empathy (their response to another's hurt is limited to their own feelings of distress), go on to respond to sanctions (eschewing bad deeds to avoid punishment), and gradually shift their focus away from the self and the immediate situation, coming to evaluate their own behavior by an external standard and finally coming to feel engaged by very general moral principles such as the sense of justice. Similarly, Bloom (2004) also describes ethical growth as an outward move from the self and the people immediately around it in concrete circumstances to universal, abstract, moral rules. The idea of the expanding moral circle is itself scalable: as we will see in the final chapters, it has been applied (if problematically) to the large arc of ethical history at a collective level. In particular, this idea plays an important role in the history of social reform and humanitarian movements, to which we will turn in the final section of the book. Before that, however, we need to look more closely at what makes ethical life something that can work at different scales. One factor is the process of making moral emotions and ethical intuitions explicit.

MAKING THINGS EXPLICIT

The process of making things explicit and therefore readily available to reflective awareness, which I will call objectification, draws on people's cognitive capacity to take a third-person perspective. But objectification is not confined to the privacy of individual minds. It depends on the existence of semiotic forms that mediate interactions *between* people. Although psychologists report that children begin to engage in "mind reading" early in life, this terminology is slightly misleading: humans are not actually telepathic. When people "read" one another's minds, or seek out intentions, interpret actions, and judge character, they are responding (consciously or, more often, unconsciously) to perceptual cues of some sort. This means that one person is construing something in his or her perceptual field to be a sign of something else. In early stages it may be immediate intentions; eventually adults draw inferences about others' character, social status, motives, attitudes, and intelligence and even their destiny, their conscience, the state of their soul, or their karma. Anything that is perceptible can be a sign. Being perceptible, signs have semiotic form, palpable qualities or features that can give rise to inferences, emotions, or even just physical actions—for when you instinctively raise your arm to ward off a blow you are treating another person's movements as signs of a pending attack. Semiotic forms may be found in noises, gestures, tone of voice, images, smells, artifacts, or any other potentially relevant aspects of one's perceptual surround. As people interact with one another, they are constantly responding to semiotic forms, mostly in quick, routine, and unself-conscious ways. Language is a preeminent source of such cues, but it is not the only one. Its cognitive and social character does, however, make it a primary vehicle for rendering concepts explicit and objectifying them. Objectifications facilitate the generalization and abstraction characteristic of a third-person perspective. Much of what follows in this book is an account of the ways that ethical categories become explicit and how they link the natural and social histories of ethical life.

At this point the reader might object that if we take seriously Haidt's account of moral dumbfounding, then descriptions and other objectifications cannot play a significant role in ethical life. They are merely post hoc rationalizations that do not accurately reflect the moral emotions that they claim to explain. As Haidt puts it, moral intuition is "the sudden appearance in awareness of a moral judgment, including an affective valence (good–bad, like–dislike), without any conscious awareness of having gone through steps of searching, weighing evidence, or inferring a conclusion" (2001: 818). But even if moral intuitions do not start in awareness, their "sudden appearance" means that they are likely to end up there: they are not only instinctive and automatic reflexes. They also are potential objects of reflection, criticism, justification, and so forth.

The question, then, is, What role does awareness play? One role, of course, is simply that of bringing one's gut reactions into one's own field of consciousness. But at least in some cases, the process seems to produce emotions, not merely draw attention to them. As noted above, Gibbard (1990: 132) argues that anger, for example, is not a moral emotion in itself. It only becomes so when it has an object, when it is anger *about* something that it makes ethical sense to be angry at, such as injustice. That quality of being "about something" is not inherent in those feelings that come to be called "anger." For one thing, you can feel anger without being able to point to a source (this experience was important for the feminist consciousness-raising groups we will meet in chapter 5). More interestingly, it seems that a given neurophysiological state may or may not give rise to the emotion of anger in the first place. Gibbard refers to an experiment by Schachter and Singer (1962) in which subjects were injected with epinephrine, which induces symptoms of emotional arousal. Some of the subjects were told that the injections would produce arousal; others were not. Each was then left with a confederate of the researchers, who with some subjects acted playful and with others acted angry. Those subjects with a playful companion felt euphoric, and those with the angry one felt angry, but in either case, the effects were more muted for those who had been informed of the drug's effects. For our purposes, this finding suggests two things. On the one hand, the actual neurophysiological effects of epinephrine, while indisputably real, are not in themselves sufficient to define the specific emotion that results. On the other hand, the particular context of being informed of those effects or not and having a playful or angry companion does not simply override the neurophysiology, which does, after all, produce actual effects in all cases. But there are several potential outcomes, so the possible effects are in a many-to-one relation to the workings of a single chemical, the epinephrine. It would seem that the relationship between physical cause and emotional effect is, at least in this instance, underdetermined (see Larsen et al. 2008). If this is right, then identifying one's feelings of arousal explicitly will have effects on the resulting emotional experience that cannot be discovered within the neurophysiological sources of those feelings alone.

Awareness can play another role in defining moral emotions. In some (but not all) moral traditions, something like revulsion at incest would be properly ethical only if you are aware that incest is revolting because it is wrong, not if you simply avoid it instinctually or, say, you are repelled due to your fear of the possible genetic risks for the offspring. Hoffman and Bloom both lay out developmental trajectories in which the ethically mature individual has transcended or supplemented immediate bodily and emotional reactions with thoughts, principles, and reasons. In this respect they are echoing a venerable tradition in moral philosophy. Kant, for example, rejected the idea that morality could be based on emotion or sentiment, much as his successors would refuse to identify it with preferences, on the grounds that emotions do not give *reasons* for moral judgments. It might turn out that the moral emotions are merely preferences, which lack the authority and sense of obligation, of rightness, that comes from justification.

But both the Kantian and utilitarian philosophical traditions demand much more than awareness; they ask for moral reasoning. Even philosophers who grant a key role to emotions, such as Gibbard, define their moral nature with reference to what it *makes sense* to feel, such as anger at injustice. Does psychological research debunk these traditions? After all, moral reasons can often be shown to serve the self through impression management or by defending a cultural worldview against challenges. Does this reduce moral reasoning to the role of a lawyer, "merely employed only to seek confirmation of preordained conclusions," as Haidt puts it (2001: 822)? A less demeaning answer is also hinted at by the same author: moral reasoning plays "a causal role in moral judgment but only when reasoning runs through other people" (2001: 819). The rest of this book, while not restoring reason to quite the same place it holds in some philosophical traditions, does seek to situate it in actual empirical contexts. It does so by taking those other people very seriously. In the process, it also considers other people not simply as the objects of the self's ethical concern or acts but as crucial to the self's ethical self-awareness in the first place.

ETHICAL AFFORDANCES IN PSYCHOLOGY

The role of other persons in the self's ethical life is the central topic of the chapters that follow. Before proceeding, however, let us review some key points from the discussion so far. One has to do with the relationship between the innate propensities and capacities that all healthy children seem to possess or come to develop and the ethical life in which they play a part.[6] On the one hand, there seems to

[6]Whether we should make "healthy" individuals the standard by which to understand moral psychology is subject to debate. For example, some activists in the neurodiversity movement argue that autism be treated as an alternative way of being a person, not a pathology.

be good evidence that certain capacities and propensities, such as empathic responses to other people's pain and pleasure, aversion to those who harm others, impulses to sharing and cooperation for their own sake, spontaneous revulsion, valuing fairness, intention-seeking, and conformity to norms, get their initial impulse from the child him- or herself. They are not, in the first instance, things that the child is taught by other people. To this limited extent, they weigh against the very strongest versions of the idea that ethical life is socially constructed right down to the fundamentals (a version that, at any rate, may exist only as a straw man for polemicists to attack). But the evidence also seems to push against any strong form of innateness or neurophysiological determinism as well. That is, we should not draw from this evidence a conclusion that humans, as individual self-contained biological organisms, are genetically predisposed to ethics or morality. One plausible way to describe this result is what has been called "starting-state nativism" (as opposed to the stronger, more teleological idea of "final-state nativism"). According to this view, the newborn is provided with "'discovery procedures' for developing adult common-sense psychology, but the final state is not specified at birth or through maturation alone" (Meltzoff and Brooks 2001: 172). What produces that "final state"? From the research discussed so far, there seem to be at least two possible answers, neither of which necessarily excludes the other.

The first answer is that, like language acquisition, these discovery procedures require input and reinforcement from other people if they are to develop at all. In one model of cognitive development, the infant's own activities play an active role in eliciting that input from caregivers who respond in

> an intensely sympathetic and highly expressive way that absorbed the attention of the infants and led to intricate, mutually regulated interchanges with turns of displaying and attending. The infant was thus proved to possess an active and immediately responsive conscious

---

For some of the paradoxes to which this position can lead, see Antze 2010; on the problem of conflating statistical averages with social or moral norms, see Canguilhem 1989.

> appreciation of the adult's communicative intentions . . . called *primary intersubjectivity.* (Trevarthen and Aiken 2001: 5)

Against pure socialization models, here the child is an active participant, seeking to engage others. Against pure biogenetic determinism, the outcome will vary depending on the nature of those other persons and their ways of responding to that engagement. The effects of that engagement later in life are the subject of the following chapters.

A second answer concerns how we identify psychological phenomena as "ethical" at all. The various capacities and propensities we have looked at here are a heterogeneous bunch. They do not necessarily fit together. One attempt to identify the neurocognitive basis for moral reason, for example, concludes that there are four systems, each dealing with a different kind of social rule processing, responsible, respectively, for care, reciprocity, social convention, and disgust (Verplaetse et al. 2009). Why should these all add up to a single thing we call "ethics" or "morality"? A similar observation applies to altruism in young children, which also seems not to be the expression of a unitary set of psychological mechanisms (Warneken and Tomasello 2009). More generally, others have observed that cognitive capacities such as executive function and recursion "are equally useful in memory, language, social cognition, and tool manufacture, and have no design features specific to the social domain" (Stone 2006: 105). Confining themselves to the apparent neurological and cognitive sources of altruism, Greene and Haidt suggest that even they are too heterogeneous to form a natural kind, concluding that "there is no specifically moral part of the brain. Every brain region discussed in this article has also been implicated in non-moral processes" (2002: 522). Notice the parallel to the conclusion that Shweder, Mahapatra, and Miller (1990) draw from the cultural evidence, remarked on above, that every apparently nonmoral event can be made moral. These researchers differ in the kinds of evidence they look at and the models with which they work, but they seem to converge on this: the phenomena we might call "ethical" cannot be identified with a single unifying basis in organic structures, cognitive functions, or cultural content. If this is correct, then what gives ethics its apparent coherence? I will argue that we need more than the basic equipment and the individual self to account for it. What pulls these diverse phenomena together is, at least in part, a set of processes that come from "above," as it were. These will include both the interaction with other persons, which is the topic of the next section of this book, and the longer-term ethical projects that life in societies gives rise to, the topic of the final section of the book.

PART TWO

Interactions

CHAPTER 2

Selves and Others

GIVING ACCOUNTS

The argument of this book so far centers on the basic claim that ethical life is not essentially something I do, feel, or cognize all by myself. Nor is it just something that I do to others. But it is also not merely the enacting of established cultural or religious norms that I simply encounter or into which I am socialized. Those who have treated ethical life in these various ways, however, have not been utterly mistaken. Each view captures some important dimension of ethics. The challenge, then, is to understand their articulations and the processes that bring them together and draw them apart. This is the first of three chapters that will focus on the details of social interaction. The purpose is to tease out the ethical dimensions of interaction and of dynamics that can prompt reflexivity and sometimes the development of more explicit systems of rules and norms. Since everyday interaction flows smoothly because it is mostly habitual and beneath the level of awareness, this section of the book asks, What induces ethical self-awareness, and what results?

When the philosopher Judith Butler asks how people come to make explicit ethical claims for themselves and their actions, she argues that they are called on to do so in the context of other people's suffering. Commenting on Nietzsche's *On the Genealogy of Morality* (1997), she writes that

> in asking whether we caused such suffering, we are being asked by an established authority not only to avow a causal link between our own actions and the suffering that follows but also to take responsibility for these actions and their effects. In this context, we find ourselves in the position of having to give an account of ourselves. (2005: 10)

Notice three aspects of this situation, as Butler describes it. First, giving an account of oneself does not occur spontaneously, nor is the speaker the only agent behind this action; there is someone other than the speaker who instigates it. This suggests that in trying to understand the language of ethical life, we cannot just focus on the content of people's accounts. It is crucial to attend to both the circumstances and the participants who summon them forth. Second, some notion of causality is involved. It follows that as ideas about causality vary across histories and societies (do local ideas about causality include divinities? witches? the will? neurons? viruses? chance? fate?), so too will both the content of the account and whether or not an account is called for. Third, there is the notion of responsibility—an avowed relationship to causality (I caused this to happen)—which implies some further consequences, such as punishment, forgiveness, or, say, a decision that no one is responsible at all.

Viewed in this Nietzschean perspective, it is an established authority that provokes account-giving, and Butler stresses the power relations involved:

> I have been addressed, even perhaps had an act attributed to me, and a certain threat of punishment backs up this interrogation. And so, in fearful response, I offer myself as an "I" and try to reconstruct my deeds, showing that the deed attributed to me was or was not, in fact, among them. I am either owning up to myself as the cause of such an action, qualifying my causative contribution, or defending myself against the attribution, perhaps locating the cause elsewhere. (2005: 11)

Her discussion does not stop there, but in these preliminary remarks, she succinctly lays out some important features of ethical life that, with qualifications, are apparent in the research we are reviewing.

In these chapters I will make the case that people's self-understanding as ethical beings is most often instigated by the very dynamics of interaction. It is those very dynamics that give rise to—indeed, may demand—explicit ethical accounts. This process may well, but not necessarily, take place across differences of power, a theme to which I return in the final chapters. I want to stress that there is nothing inherent about people's judgments as such that requires them to be fully self-aware about their ethics or able to verbalize it. Note as well that, in Butler's version at least, self-awareness has a retrospective and reconstructive character. But it is important that people do (sometimes) become ethically self-aware and verbal and do (sometimes) project themselves forward in time as ethical persons—and that is crucial to the ways in which psychology and social history feed into one another. What are the empirical conditions for ethical self-awareness? Do they involve self-discovery too?

Butler's insights are developed through the exegesis of other philosophical texts in the European scholarly tradition. But if our task is to gain an empirical understanding across the full range of human experience, we need to look at other kinds of evidence. So far we have looked at laboratory experiments and observations in psychology, mostly in Western settings. In this and the next two chapters, we turn to research by linguistic and cultural anthropologists, as well as sociologists and political scientists, that looks closely at ordinary social interactions, the stuff of playground taunts, village gossip, and roommates' quarrels, as well as the teaching of politeness and exhortations to good sportsmanship. This research offers a granular view of activities that can seem quite banal. But their power lies precisely in their ubiquity: saturating the ongoing flow of experience, their ethical implications can easily seem to be simply the way things are. The flow of experience here runs amid other people. As the sociologist Harvey Sacks remarks, "Human history *proper* begins with the awareness by Adam and Eve that they are observables. . . . By the term 'being an observable' I mean having, and being aware of having, an appearance that permits warrantable inferences about one's moral character" (1972: 281, 333 n. 1). To become aware that you are "an observable," in Sacks's distinctive turn of phrase, is to find yourself amid other people, imagining their perspective on you.

INTERSUBJECTIVITY

As noted in the previous chapter, psychological research on moral dumbfounding, situational cues, and the emotional basis of supposedly rational decisions can have strong debunking effects. Findings such as these seem to undermine any sense people might have that they are masters of their own ethical lives or have real insight into others'. The discrepancy is due not just to the obvious fact that people's words are poor guides to their motives and thoughts. That could be due merely to the inherent inadequacies of language or the self-deceptions familiar to everyone. The stronger claim is that people in their very nature as biopsychological beings can have little insight into themselves as ethical subjects. In that view, humans are subject to mechanisms, causes, or correlations whose existence does not depend on their awareness of them. And yet people carry on as if they were self-motivated, self-aware, and self-interpreting creatures; moreover, most of the time they take the people around them to be so as well. So their responses to one another cannot be understood without some account of that self-awareness, both their own and what they impute to others.

As noted in the introduction, empirical research on ethics has prompted worries that the results will eliminate the dimensions of freedom and responsibility at the heart of many definitions of ethics (Laidlaw 2014). The focus on interaction that I develop in these central chapters of the book seeks to address this problem from a somewhat oblique angle. It does not directly reassert a particular concept of freedom. Rather, it explores a crucial feature of interaction, that the outcome is never fully in anyone's hands and that each actor responds not just to what the others do but to the available means of making sense of what they do and evaluating it. When people try to claim or deny responsibility for their actions, they often do so by defining those actions in ways that will get others to assess them in certain ways and not others (cf. Austin 1979: 180). We will return to this point in chapter 4.

The commitment to taking people's self-understanding seriously matters because it plays a crucial role in ordinary social existence. Whether or not people accept one another's self-understandings at face value, they must at least respond to them in some way, even if only with suspicion. In this chapter, I will approach the problem of self-understanding by looking at research on intersubjectivity and interaction. I will argue that anything that can be described in ethical terms involves people's interactions with one another. Moreover,

I will try to show that interactions are *not* best understood as the encounter between autonomous individuals. But my purpose is not to deny or ignore the perspectives of the sciences; rather, it is to gain a better understanding of how they fit with the world as people experience it.

One point of convergence between psychological and social perspectives on ethical life is *intersubjectivity*, people's capacity to share and exchange perspectives and intentions with one another. Viewed in developmental terms, as noted in the previous chapter, reciprocity of perspectives is evident early in childhood, when a child points to things to direct the attention of another person or coordinates his or her gaze with another's (Hobson 2004; Tomasello et al. 2005). Acts such as these seem to depend on recognizing that the other person's point of view is not the same as the self's and that the two can nonetheless be brought into coordination or placed in opposition. Otherwise there is nothing to be reciprocal about. But to know that what I see is not what another person sees is a developmental achievement. Overcoming that difference through cooperation and coordination goes beyond direct empathy, since it depends on the child's ability to identify with others insofar as they *are* others, as someone not identical to him- or herself.

The sequence of events is revealing: for children the developmental challenge is as much a matter of learning to be separate from others as it is one of learning to get along *with* them. At least in some accounts, the child must first learn to differentiate others from the self, before being able to engage in such reciprocity. For example, Tomasello (1999: 181) refers to an experimental finding that young children sometimes recall having done an action that was really done by another. Findings such as these suggest that the development of social skills is not simply a matter of overcoming a primordial difference between discrete individuals but, rather, the emergence of a sense of self from an initial condition lacking distinctiveness (Ochs 1988; Schieffelin 1990). This runs against the assumption built into some evolutionary models that take their challenge to explain how natural selection could form society starting from something like game theory's "egoists without central authority" (Axelrod 1984: viii).

According to Michael Tomasello,

> To comprehend the linguistic communications of others, children must in some sense simulate the perspective of other persons as they are expressing themselves linguistically, and so the back-and-forth of discourse involves the child in a constant shifting of perspectives from her own to those of others and back again. (1999: 176)

Once they begin to master this back-and-forth, their facility in doing so is aided by linguistic devices such as the first- and second-person pronouns, elements that no language can do without. As the linguist Emile Benveniste notes,

> I use *I* only when I am speaking to someone who will be a *you* in my address. It is this condition of dialogue that is constitutive of *person*, for it implies that reciprocally *I* becomes *you* in the address of the one who in turn designates himself as *I*. (1971: 224–25; cf. Hill and Irvine 1992; Rumsey 2003; Trevarthen and Aitken 2001)

Children must already be able to enter into reciprocity of perspectives before they can speak. But things do not stop there: basic intersubjectivity may be a condition for entering into language, but once the child can speak, language becomes a necessary condition for further developments in that child's intersubjective abilities. As we will see, this kind of bootstrapping is characteristic of the role that objectifications can play in ethical life.

Semiotic mediation plays a critical role in that development. As noted above, since people are not telepathic, even their most intuitive, intimate, and insightful construal of others is always mediated semiotically (Deacon 1997; Enfield 2013; Keane 2003, 2010; Lee 1997; Silverstein 2003a). What this means, in the case of social interaction, is that people draw inferences about one another from perceptual cues or signs. Semiotics need not refer to conventional codes or purposeful signals—what I might infer about your feelings or intentions, for instance, may be idiosyncratic or may exceed anything you mean to reveal or are even aware of. Language, of course, plays a central but not exclusive role in this process, which, in principle, can draw on anything at all that can be perceived, such as voice timbre, bodily posture, smell, or even the physical location or time of day. Although the point may seem obvious, the implications are not trivial; as later chapters will show, semiotic form mediates between the natural and social histories of ethics.

First, however, we need to look at the ethical dimensions of interaction as such. For, as discussed in chapter 1, it is sometimes claimed that joint attention and reciprocity of perspectives naturally lead to ethical life. But neither joint attention (for instance, child and parent knowing that they are both looking at the same tool together) nor reciprocity of perspectives (for instance, a child learning to see herself from the father's point of view) determines or necessitates the emergence of ethics. Rather, they are best understood as affordances that ethical judgments and practices take up and develop. To make sense of this we need to consider two aspects of the interactions, the imputing of intentions to one person by another and the sense of shared reality. Both are commonly taken to bear on the tenor of people's relations with one another; both contribute to people's imputations of motives and character to one another. They involve different features of interaction, however, which I will discuss in turn.

INTENTION-SEEKING

In the previous chapter, I argued that ethical life cannot be based solely on empathy, because empathy itself might just be egocentric, a response to another's pain as if it were mine. A more full-fledged ethical response would require that I have some fuller appreciation of another person's independent existence in a way that matters. An important step in that direction would be to take others to be people who have intentions, goals, desires, and values, much as I do, but ones that are not my own. Reciprocity of perspectives is a step in that direction. Semiotic devices, notably but not exclusively linguistic, facilitate the development of this reciprocity. In particular, pragmatics, how language is used, seems both to build on and to contribute to the perception that other persons have potentially discernible intentions. As noted above, people are not exactly mind readers; rather, they are sign readers. This is a crucial point to which I will return again and again, because it shows us a critical point of articulation between psychological and historical processes. But first we must look more closely at intentions and the dynamics of interaction.

As we saw in chapter 1, infants become oriented to other people's intentions, and even evaluate them, very early in life (Hamlin, Wynn, and Bloom 2007; Meltzoff 1995; Onishi and Baillargeon 2005; Tomasello 1999; Woodward 1998). Building on reciprocity of perspectives, the capacity to infer another person's likely intentions is a component of so-called Theory of Mind (Baron-Cohen 1995; Wellman 1992; Wellman and Liu 2004). Among the diagnostic features of Theory of Mind, over the course of development, are the abilities to distinguish between intentional and unintentional actions, classify actions according to goals, direct joint attention with another, enter into pretend play and deception, and grasp that mental states are independent of physical reality. Some ethnographic research has been said to raise doubts about the universality of this view of intentionality, on the basis of claims that people in some societies say that they are unable to interpret other persons' intentions. In chapter 3, I will return to that question and suggest another way to understand the evidence. But first let's take a look at the role of intention-seeking in verbal interaction.

The propensity for intention-seeking, which has clear sources in the natural history of humans, may afford ethical reflection and reasoning. Certainly the philosopher P. F. Strawson thought it to be a precondition for some kinds of ethical evaluation when he wrote,

> If someone treads on my hand accidentally . . . the pain may be no less acute than if he treads on it in contemptuous disregard of my existence. . . . But I shall generally feel in the second case a kind and degree of resentment that I shall not feel in the first. (1974: 5)

Notice the double reflexivity this resentment requires: I must both impute intentions to another and assume that the other anticipates that my own feelings, and not just my hand, will be hurt as a result. Although the ubiquity of human intention-seeking may have neurological or cognitive sources, its role in ethical judgment is not itself deterministic. Rather, as with the details of interaction discussed above, it seems to be an affordance to which people may respond in the process of making ethical judgments.

In the view that Strawson expresses, intentionality lies at the heart of an ethical response to interaction. Or more precisely, perhaps, what is necessary is being able to construe acts, events, or words as *signs* of intentions. As I learned during my fieldwork on Sumba, much depends on what one takes to be a plausible sign of a plausible intention. Sumbanese are prone to a kind of hyperhermeneutics in which there are no accidents. This leads them to seek intentions even where I would have found them unlikely. This was apparent in one situation I have described elsewhere (Keane 1997: 29–33). After a tense misunderstanding, one man presented his brother-in-law with a fine but damaged textile, carefully folded in formal gift mode. Gifts of this sort in Sumba carry enormous ceremonial and political weight. But when the cloth was opened up, the recipient discovered that there was, in fact, only half a piece. Unbeknownst to the giver, the rest had been accidently torn off earlier in the day. But the recipient refused to believe this and took the gift as a serious insult. This was a wholly characteristic response in a world where a stumble, a fit, a shortage of betel nut, and a sudden windstorm or lightning strike are all probed for dark intentions harbored by agents visible or invisible. Such ways of making sense of events as signs treat the world as saturated with ethical import (see Keane 2014).[1]

Language provides an especially subtle and pervasive source of such signs. As noted above, being able to use language depends on a basic capacity for reciprocity of perspectives. But the role of intersubjectivity goes beyond taking the other person's perspective. For one thing, language is not simply a code or set of rules governing the transmission of messages from one person to another, and understanding language is not merely a matter of decoding those signals. In actual use, the flexibility and power of language depend, in part, on the role played by inferences. Not all inferences point to intentions (one can infer someone's hometown from their accent, for example, their education from their vocabulary, or their emotional

[1]There are many instances of the opposite too, a paucity of intentions with a surfeit of ethical responsibility. One example is what Bernard Williams (1981) calls "moral luck." Oedipus accepts punishment for a crime no one thinks he intended. A safe truck driver understandably feels guilt for the death of a child who ran in front of his vehicle.

state from their tone of voice), but many do. To the extent that language use does presuppose intentions on the part of other people, it helps weave into the fabric of everyday interaction the notion that other people are (at least potentially) ethical subjects with intentions, goals, and values. The way people draw inferences from each other's speech is, implicitly, to treat others as if they were creatures with intentions, goals, desires, and values—thus, at least potentially, as ethical subjects. (This does not preclude the possibility that they may treat some humans as not fully ethical subjects, like slaves to be commanded or foreigners whose language cannot be understood, or that nonhumans might not also be included as ethical subjects on grounds other than the dynamics of language use, as the Kluane treat rabbits. These possibilities are historically pliable, as we will see in the final set of chapters.)

CONVERSATIONAL INFERENCES

One influential effort to formalize the workings of inference in language use is that developed by the philosopher H. P. Grice. Consider irony or hints, conversational moves that cannot be captured by any set of rules or codes. To make sense of how they work, we need to consider the intentions and inferential abilities of both speaker and addressee. Grice introduced two concepts to try to capture this. One is "nonnatural meaning." He summarizes it this way: "*A* uttered *x* with the intention of inducing a belief by means of the recognition of this intention" (1957: 384). In contrast to "natural meanings" like "smoke means fire," nonnatural meanings require reference both to the speaker's intentions and to the addressee's ability to recognize them. Notice the parallel to Strawson's example of having one's hand stepped on by someone else. If someone treads on my hand by accident, I may suffer physical pain, but I should feel no resentment. But it is an insult if the person stepping on me wants me to know that it was done on purpose, in order to make me feel insulted.

Grice's second concept has to do with how speakers can prompt listeners to draw inferences from their words. Things such as hints and irony work through stylistic variations that invite the listener to construe intentions without making them explicit in the semantic content of the words spoken. To formalize this, Grice (1975) proposed that everyday talk works on the basis of "maxims." The maxims are heuristics, default assumptions that help people interpret actual speech. Grice sorted them into four types: quality (e.g., do not say what you think is false), quantity (e.g., do not say more than necessary), relation (be relevant), and manner (e.g., avoid obscurity). Maxims are not rules that people follow but a background against which certain moves become noticeable. When a speaker violates a maxim, he or she is inviting the listener to seek out a reason why. For example, if an irate mother addresses her son, normally called "Wig," with his full name, "Eldridge Wigglesworth Potemkin Junior!" she is violating the maxim of quantity and inviting an inference about her state of mind.

Here's an empirical example from a recording made by a conversation analyst. The speaker is a waitress in a British pub complaining about a friend, who was below the legal drinking age, trying to get away with ordering alcohol during her shift:

> So I walked up to Cathy en I said OH, yer drinkin a Coke, can I have a sip? So I took a sip of it, EIGHTY PROOF. I couldn't believe it. Jimmy pours Bacardi and Coke. She had some guy go get it for her. So I looked, I took the drink, and I drank out of it. And I looked at her. (Drew 1998: 308–9; transcription simplified)

Violating the maxim of quantity, the overdescription of the action of taking a sip (I looked, I took the drink, I drank out of it, I looked) emphasizes the speaker's volition to dramatize the moment of moral outrage. Another example also shows how the maxims can be used to stress another person's intentionality in order to pin responsibility on him or her. Here a trial lawyer overdescribes an action in order to stress the agent's deliberateness and therefore ethical responsibility, with the words "So the defendant took the car and drove it into some trees, didn't he?" instead of "The defendant drove into some trees" (Drew 1998: 318). These small stylistic options draw attention to themselves in order to lead the listener to certain conclusions. Moreover, in both these cases they do so in order to carry out some ethical work: by emphasizing the intentionality of the transgressive action, they are effectively making an accusation against the actor.

SHARED REALITY

In Grice's formalization, conversation is based on an underlying "principle of cooperation" that generates such maxims as "Be truthful." Grice does not seem to have intended any specifically ethical interpretation of the principle of cooperation; it could, for example, be understood merely as a technical device to make communication more efficient. As we will see, even a bitter quarrel requires cooperation, since it depends on the participants agreeing that they are quarreling—an agreement that is not always guaranteed.

Not surprisingly, philosophical attempts such as those of Grice have been criticized by anthropologists for relying on assumptions about truthfulness that, for example, are either highly local in themselves or require local specification in order to function (Hanks 1996; Keane 1997, 2007; Keenan 1976; Rosaldo 1982; Sweetser 1987). But linguistic evidence suggests that the fact of variation in default assumptions as such hardly disqualifies Grice's major insight, that social interaction depends on inferences, which in turn draw on background assumptions to trigger them. Moreover, even apparently instrumental features of interaction—means toward communicative ends—may be inseparable from human propensities to make value judgments on the basis of semiotic forms (e.g., "quantity" of speech) in ways that cannot be reduced to the technical requirements of communication. In short, whatever background assumptions people may bring to bear in their interactions can afford ethical judgments.

To illustrate the rhythm of an everyday quarrel, linguistic anthropologist John Haviland analyzes a recording of two angry roommates (here I give just the translation from the original Spanish). One, who has recently returned from an emotionally difficult trip, is offended that the other has not asked her about it:

22 **WE HAVEN'T SPOKEN** EVEN ONE NIGHT
23 L; I FEEL THAT **I HAVE SPOKEN** WITH YOU
24. P; **YOU HAVE SPOKEN** WITH ME BUT ABOUT YOUR TROUBLES
25 L; I FEEL THAT **I HAVE SPOKEN** WITH YOU
26 P; JUST ONE SINGLE TIME

27 YOU HAVE HAD TEN MINUTES **TO SPEAK WITH ME** (HAVILAND 1997: 561)

Haviland shows that they agree on what is happening and shape their speech accordingly. Thus, in the quarrel between roommates, as the hostilities build up, the two "toss a kind of phrasal ball back and forth, each using the appropriately inflected frame *haber hablado con* __ 'I have talked with __' [translation transcribed in boldface above] in an unresolved game of catch-the-blame" (Haviland 1997: 561). Their agreement on how to argue is manifested in this jointly constructed rhetorical structure.

The intersubjectivity on which the arguments depend is constructed, in part, by linguistic and other semiotic forms that enter into the participants' evaluations of one another and one another's actions. Without this degree of intersubjectivity, people cannot even successfully quarrel with each other. For example, anthropologist Ilana Gershon, carrying out research on the romantic lives of American undergraduates, quotes one student describing a conversation with her boyfriend:

> I got into an argument, I guess we got into a fight and I didn't even know that we got into a fight, I thought we were just arguing. . . . I thought we were just having a discussion. And he was really mad and he strode off. And I went "wait, are we in a fight?" (2010: 400)

As Gershon remarks,

> Breakups are often confusing moments, and one is not always certain a breakup is taking place, even if you are the one initiating the breakup. Frank talked about how he had tried to breakup several times with Maria, but the breakup conversations he initiated never seemed to result in a breakup. (2010: 396)

Even a breakup requires some cooperation and agreement on the nature of the action that is under way.

Cooperation in this sense is a technical requirement for an interaction to proceed. But it can be more than that. Like successful quarrels, interactions depend on the participants constructing a shared sense of the reality of the immediate situation. They agree on what is going on and, given that, what the appropriate way to act is. Banal departures from appropriateness can be revealing. As the literary critic Michael Warner remarks, "You could be gossiping in a corner and suddenly realize that everyone in the room is listening. You could be practicing a formal speech at home only to discover that it sounds pompous and corny" (1999: ix). The language of the absurd and the pompous suggests how apparently trivial matters of style may, under the right circumstances and given sufficient reinforcement, give rise to stronger judgments about a person's character. Such judgments are the everyday workings of the often unwitting exclusions of class, ethnicity, and gender. As anthropologist Philippe Bourgois observes of the streetwise Puerto Ricans with whom he worked in New York's El Barrio, "In the high-rise office buildings of midtown Manhattan or Wall Street, newly employed inner-city high school dropouts suddenly realize they look like idiotic buffoons to the men and women for whom they work" (2003: 143). These were men who might be enormously competent in other contexts, like this drug dealer:

> Ray's mastery of street culture enabled him to administer his businesses effectively in the drug economy. He skillfully disciplined his workers and gauged the needs of his customers. He mobilized violence, coercion, and friendship in a delicate balance that earned him consistent profits and guaranteed him a badge of respect on the street. In contrast, in his forays into the legal economy, Ray's same street skills made him appear to be an incompetent, gruff, illiterate, urban jíbaro to the inspectors, clerks, and petty officials who allocate permits and inventory product, and who supervise licensing in New York City. (Bourgois 2003: 135)

But we do not need to look for gross differences of ethnicity, class, or gender to encounter the role of evaluation in minor aspects of interaction. In a famous experiment, the sociologist Harold Garfinkel had his students spend fifteen minutes to an hour at home with their parents, acting as if they were boarders. In response to the small changes in their interactions, such as extra politeness, the parents commonly reacted in strong terms:

> Reports were filled with accounts of astonishment, bewilderment, shock, anxiety, embarrassment and anger, and with charges by various family members that the student was mean, inconsiderate, nasty or impolite. Family members demanded explanations: What's the matter? What has got into you? Did you get fired? Are you sick? What are you being so superior about? Why are you mad? Are you out of your mind or just stupid? One student acutely embarrassed his mother in front of her friend by asking if she minded if he had a snack from the refrigerator. "Mind if you have a little snack? You've been eating little snacks around here for years without asking me. What's gotten into you?" One mother, infuriated when her daughter spoke to her only when she was spoken to, began to shriek in angry denunciation of the daughter for her disrespect and insubordination and refused to be calmed by the student's sister. A father berated his daughter for being insufficiently concerned for the welfare of others and for acting like a spoiled child. (Garfinkel 1967: 47–48)

Here mere matters of appropriateness that undermined the sense that participants agreed on the context and the nature of their relationships were deeply unsettling and prompted judgments about character and other aspects of how people ought to value one another.

How do we understand the remarkable animus of the parents' responses to Garfinkel's students? The focus on intentionality I have discussed so far does not seem entirely to answer this. The problem is not merely a matter of messages gone astray. We need to consider as well the ways in which interactions build up a sense of shared reality and establish people's regard for one another. In fact, these are often interconnected. Conversational interaction depends on a basic capacity for intersubjectivity. But intersubjectivity cannot simply be assumed. It is constructed by means of a repertoire of things that are perceptible, such as tone of voice, winks, and even silence, which afford inferences about what they might be signs *of*.

The sociologist Emanuel Schegloff remarks that interactions depend on "methods for designing talk to do recognizable actions and methods for recognizing the actions so designed by coparticipants"

(2006: 87). To be recognizable, an action must have a design, that is, certain tangible properties that make it so. For conversation analysts, these properties are structures of turn construction, action sequencing, and repair, forming "an 'architecture of intersubjectivity' in which previous, current and next components of a sequential organization interlock and reinforce one another" (Sidnell 2014: 364; cf. Atkinson and Heritage 1984; Enfield 2013). Any given moment in a conversation, one person is interpreting what the other person has said. By continually checking, revising, and rechecking one another over time, people draw on semiotic forms to build up a sense of shared implicit description of who they are and what they are doing: that they are joking or serious, quarreling or engaging in banter, gossiping or making plans; that they are friends, business partners, lovers, fellow passengers, bosses, servants, rivals, and so forth. This sense of shared reality is crucial to what Garfinkel (1963) called "trust," namely, the confidence that we are interacting with people who are committed to the same definitions of reality to which we adhere. Notice that the process of responding to the other person over the time span of the interaction means that the shared reality is not simply a matter of following a template or schema; it is emergent and subject to reevaluation over the course of the interaction. What the parents in Garfinkel's experiment display is one possible consequence of challenges to that sense of reality: they seem to draw strong *ethical* inferences from apparently superficial *forms* of behavior. For our purposes, two things are important in this approach to interaction: those forms of behavior characteristically anticipate the perspective of the other person, and their realization as intersubjectivity depends on the response and possible reframing by that other person over a stretch of time. If ethical life depends on intersubjectivity, it cannot be just a result of empathy or some other *individual* psychological disposition. First, it is built up, reshaped, or undermined, in time, between people; and second, this is done by means of perceptible material forms such as language, bodily deportment, and so forth. As historical circumstances change, the resulting sense of reality may become unacceptable to some or all of the participants. We will return to the topic of descriptions in chapters 4 and 5; the final set of chapters will show some of the ways in which seemingly minor elements of interaction can become targets of social critics, reformers, and revolutionaries.

REGARD FOR ONE ANOTHER

Whether or not the Gricean maxims successfully capture some universal principles of conversation equally applicable from Samoa to Oxford, they do help us understand the role that default assumptions (however local they might be) play in provoking people to draw inferences about one another's intentions, the nature of the action taking place, and one another's character. Commitment to the expectations of interaction seems to blend easily into commitment to the other person (Clark 2006). As Garfinkel's experiment shows, one person's neglect of the technical demands of routine joint activity might be taken by others as a failure of social recognition, an insulting withdrawal from one's expected role in securing the other's sense of self.

Normally people are unaware of these normative expectations until the question of whether they will be fulfilled arises over the course of interaction. It is on the basis of those interactions that people draw conclusions about one another and their relationships. As the linguistic anthropologist Jack Sidnell puts it:

> Interaction is itself a moral and ethical domain. When persons interact they necessarily and unavoidably assess whether they are being heard, ignored, disattended, and so on. Is this person really listening to me? Paying attention to me? And, in so doing, acknowledging me as a worthwhile person who merits such attention? (2010b: 124–25)

Now "being heard, ignored, disattended, and so on" could be construed as merely a technical problem with channels of communication, like testing whether a microphone is turned on. But the point is that, by virtue of the inferential capacities identified with Theory of Mind, attentiveness and inattentiveness register, or at least are available for being taken to register, people's regard for one another.

Consider the practices of "repair." This is a ubiquitous feature of conversation by which participants try to resolve what they perceive to be problems in the interaction itself. Often these problems are about technical glitches, such as confusion in turn-taking, or ambiguity about how a certain remark is to be taken (is it serious or facetious? an information question or an indirect request?). But according to Schegloff, the main focus of repair is intersubjectivity. Repairs arise when people perceive themselves to be losing "the shared reality of the moment" (Schegloff 2006: 78) and act to restore it. Moreover, Schegloff continues, we should not assume that intersubjectivity is normally stable or given. Rather, "practices of repair and their ordered deployment are probably the main guarantors of intersubjectivity and common ground in interaction" (Schegloff 2006: 79). Thus, although the capacity for reciprocity of perspectives and shared frames or points of view may be cognitively very basic to all humans, in any given instance they require perceptible media to sustain. Those media arrive bearing the marks of their distinctive histories.

Repair is not just a technical matter. This is the point of Garfinkel's experiment with his students and their parents: minor differences of demeanor threaten the shared sense of reality within the family. The result is not merely a lack of coordination but a stronger sense of threat and hurt. There is a continuum from the minor technical glitch—two people starting to speak at the same moment—to the embarrassment of speaking in the wrong tone that Warner describes, to the shaming, humiliation, and potential job loss experienced by men of El Barrio in midtown office buildings; the stigmatization of social outcasts; and even the exclusion of some categories of people from humanity altogether.

The crucial point here is that judgments saturate interaction and take its resources—which serve a diversity of communicative functions—as ethical affordances. Between the extremes of minor glitches in the conversational flow and outright racism, sexism, and the like are situations such as the following, arising from the most ordinary of service encounters, ordering coffee at Starbucks. As linguistic anthropologist Paul Manning analyzes these encounters, they can become fraught due to contradictions that disrupt the smooth flow of interaction. Starbucks projects an aura of high status, due to the complicated, and to many people, unfamiliar, taxonomy of drinks on offer. Problems arise when baristas correct customers who fail to order drinks using the correct terms in the right sequence. From the point of view of the barista, proper ordering is simply a technical matter of making the job go more smoothly. But the correction is what conversation analysts call a potentially threatening "other-initiated repair" of the customer's speech. It can seem to customers to be an assault. Some customers seem to be predisposed to this sense of vulnerability because of contradictions built into the situation. The contradictions are multiple: baristas have expert knowledge of the taxonomy but are also structurally subordinate to customers (who are supposed to be "always right"). This contradiction of hierarchies is crosscut by a basic egalitarian ethos governing interactions in middle-class American life that can make the fundamentally unequal nature of service encounters the locus of heightened sensitivities all around. Studying a Web site where Starbucks baristas could vent after an irritating day at work, Manning found that these sensitivities were expressed in their rants about bad customers:

> Because of the way that Starbucks overlays class anxieties . . . on an already fraught customer–server relationship, some customers treat the attempt at repair to be in itself a face-threatening act of "correction," or will obstinately refuse to cooperate, or will continue to blunder forward in confusion, leading the conversation to a place where the issue is no longer a technical crisis, but a normative one. At such points the most explicit statements of presuppositions about the hierarchical nature of the service relationship will be found, attempts will be made by customers to achieve by stipulation the respect it is felt are due all customers at the expense of the respect which is generally not felt to be due the server. The recrudescence of the aristocratic memory that haunts all service interactions is the focus here. The basic claim that is being made is that servers in service interactions also are owed the courtesy that is normatively accorded customers in general. . . . Each of these transcripts of rants about customers is haunted by the normative, but absent, image of ordinary talk between equals. (2008: 123–24)

Viewed as a problem in ethical life, these encounters reveal an endemic contradiction between a norm of equality that has deep historical roots in a larger American public sphere and equally powerful norms surrounding anonymous market interactions in consumer society. What could be described in the imposing terms of political culture or moral philosophy here come to a head in ways so trivial as to seem merely irritating or bewildering to the participants—Why was that barista so rude? Why are customers such jerks? But it is their very ubiquity and opacity that can give interactions like these their power. They seem merely to reflect reality, the natural qualities of people, in an ethical light. They play out not as conflicts about the technical apparatus of conversation or about normative life in the abstract but as judgments about people's character.

A SEMIOTICS OF CHARACTER

In the previous chapter I sketched out some of the challenges to virtue ethics that have come out of experimental psychology. To summarize, at the center of most versions of virtue ethics is the idea of character, "set dispositions to behave systematically in one way rather than another, to lead one particular kind of life" (MacIntyre 2007: 38). Yet this ideal seems to invite refutation by naturalistic explanations of behavior. Some psychologists claim, as the philosopher Maria Merritt puts it, that "few if any persons display the integrated structure of robust personality traits that would conform to the Aristotelian ideal of virtuous character" (2000: 367; see Doris 2002; Ross 1977). In response, Merritt points out that people "normally witness episodes of behavior in a quite restricted range of circumstances, in which common situational regularities recur over and over again" (2000: 373). In other words, the daily routine offers fewer of those variables that might disrupt one's sense of consistency than do the experimental protocols meant to undermine it.

The most important of the situational regularities to which Merritt refers is "the sustaining social contribution to character" (2000: 374), that is, particular interactions and settings. Even if individuals are not wholly endowed with some "internal" stability, they are supported by consistent "external" frames of reference such as etiquette, routines, institutions, and creeds, as well as reputations. To capture how these frames of reference work (at least in North American contexts), the sociologist Erving Goffman developed the concept of "face" from the English-language appropriation of a Chinese concept.[2] Once a person's character is more or less established, he argued, he or she can expect to be sustained in a particular face and "feel that it is morally proper that this should be so" (1963: 7)—other people will tend to act *as if* you are acting in character even if you are not. Indeed, potential face-threatening failure is one of the main things repair should mend. What Goffman calls "face-work" is not just a matter of inner psychological states but invokes the larger public world in which people make moral claims about themselves whose success in the first instance depends on being perceptible and recognizable as such to other people.

One familiar form such claims can take is reputation. The enduring effects of one's reputation depend on the words of others, which can in turn shape one's ongoing interactive possibilities. In his study of gossip in the Mayan village of Zinacantan, John Haviland says that it

> relies on reputation while at the same time it expands on it. The fact that a man is known as a murderer certainly affects the way others treat him and the distance they maintain, and a story about a past murderer may easily trade on the man's reputation without explicitly stating it. (1977: 58–59)

Although I do not want to claim that reputation is identical to character, such regularities of interaction do help shape the public exoskeleton of character—they form the context that can help support consistency where merely psychological factors do not.[3] As we will see in the next chapter, they are also part of the conversational infrastructure on which ethical reflexivity depends.

The ways people evaluate one another are not simply matters of applying rules or of deciphering codes, but they do depend on perceptual forms. As Goffman notes, the Greek word *stigma* referred to

[2]Goffman takes advantage of the way English usage conflates two Chinese concepts translated as "face": *Mien-tzu* emphasizes social status; *lien*, "the confidence of society in ego's moral character" (Ho 1976: 867). I thank David Porter for clarifying this.

[3]These social dynamics may work in coordination with what psychologists call the "fundamental attribution error," a tendency to explain someone's actions in terms of traits rather than situations (Ross 1977). Note that this bias is stronger in cultures that emphasize consistency of character (Gopnik, Seiver, and Buchsbaum 2013), a point we return to in chapter 3.

> bodily signs designed to expose something unusual and bad about the moral status of the signifier. The signs were cut or burnt into the body and advertised that the bearer was a slave, a criminal, or a traitor—a blemished person, ritually polluted, to be avoided. . . . Later, in Christian times, two layers of metaphor were added to the term: the first referred to bodily signs of holy grace that took the form of eruptive blossoms on the skin; the second, a medical allusion to this religious allusion, referred to bodily signs of this physical disorder. (1963: 1)

The role of the body in ethical evaluation is already apparent in the most important source for contemporary virtue ethics, Aristotle's *Nicomachean Ethics* (1941), as in this passage:

> A slow step is thought proper to the proud man, a deep voice, and a level utterance; for the man who takes few things seriously is not likely to be hurried, nor the man who thinks nothing great to be excited, while a shrill voice and a rapid gait are the results of hurry and excitement. (1125a, 11–16)

Although Aristotle may not have been claiming that the character of the proud man is identical with his demeanor, this passage does exemplify how the embodiment of virtues enters into a semiotics of value. The sociologist Pierre Bourdieu similarly portrays the way bodily habits can be (and not just express) a virtue:

> A [Kabyle] woman is expected to walk with a slight stoop, looking down. . . . [T]he specifically feminine virtue, *lahia*, modesty, restraint, reserve, orients the whole female body downwards, towards the ground, the inside, the house, whereas male excellence, *nif*, is asserted in movement upwards, outwards, towards other men. (1977: 94)

Note three things about this description. First, there is a causal relationship between social order and bodily demeanor (think how much disciplining by other people is packed into the passive verb form "is expected to"). Second, by virtue of its embodied character, this relationship is not dependent on fully cognized execution—in fact, according to Bourdieu, it is most effective when the individual is least self-conscious or reflective about it. But, third, virtue is—and *must* be—materialized in sensible forms.

Being necessarily perceptible, these materializations can be recontextualized. This can produce more conscious reflection. Transfer this Kabyle woman from Algeria to a *banlieue* in France, and the taken-for-granted signs of her virtue become the objects of highly explicit forms of ethical talk, from morally outraged condemnations by feminists or nationalists to neoorthodox justifications by clerics (Bowen 2010; Mahmood 2005). Across the spectrum from the gestures to words, perceptions of semiotic form offer a platform on which self-aware ethical reflection can, potentially, be constructed. Ethical evaluations can draw on everything from bodies and conversational patterns to doctrinal texts and philosophical logic and, in doing so, pull together quite diverse strands of natural and social history.

ETHICAL VULNERABILITY

One consequence of its fundamental sociality is that the self is vulnerable. The risks of ordinary social life are not always noticed in some of the more eudaemonistic approaches in moral psychology. Goffman, however, being ever sensitive to the dark side of sociality, remarks, "When a person volunteers a statement or message . . . he commits himself and those he addresses, and in a sense places everyone present in jeopardy" (1967: 37). Much of the developmental literature introduced in the previous chapter seeks the sources of ethics in universal aspects of human cognitive and emotional capacities. They point to apparently altruistic impulses that appear very early in child development, such as empathic responses to others and the tendency to share. These findings suggest that humans have very basic propensities for alignment with others and reciprocity of perspectives. But ethical life does not emerge automatically from the individual; it requires ongoing social interactions to elicit and reshape that individual as he or she matures. That means that any account of ethical development that relies only on the psychological predispositions of individuals will miss some crucial features. Attending to those features will help us understand the historicity of ethics, as we will see below. In this chapter and the next, we will remain in the more intimate scale of face-to-face interaction.

The term *ethical life* refers to the saturation of social existence with evaluations of persons, their relations, and their actions. As noted earlier, the scholar should be wary of defining "ethics" simply in terms of values that he or she happens to agree with. Given the complex histories of human communities, we should not assume in advance that their ethical worlds can be fully reconciled with one another or even, in any given instance, remain stable and internally consistent. On first glance, at least, there will always be some values some people hold that others will find repugnant (Doris and Plakias 2008). In the introduction, we saw that Sally's sense of justice directly contradicts moral values of a certain church. Similarly, the value of self-determination asserted by abortion rights advocates, on the one hand, and that of life asserted by their opponents, on the other, are both ethical positions. Norm-seeking and empathetic tolerance can be uneasy companions. The virtues of forthrightness and of humility are both ethical; so too the often contradictory values of loyalty and freedom (see Graham, Haidt, and Nosek 2009). We will return to this issue in later chapters, about moral revolutions. Here, I want to look at the vulnerabilities that face-to-face interaction entails. If the ethical self depends in part on the cooperation of others, as the research reviewed here suggests, then it is inherently subject to the exigencies of interaction as well. One is engaged with others' evaluations, something over which one has limited control. Indeed, the limits on individual control may even be one condition for ethics, because they make one's involvement with others an unavoidable condition for being who one is. If ethical life is not best defined by the obligations and punishments that Williams sees in the morality system, it is still marked by constraints. They come from the inescapable role of other people in any individual's ethical life.

Ethical evaluations can take advantage of the ways in which technical devices of conversational interaction and coordination seem to involve social commitments. The linguist Charles Goodwin says that one may carry out an action

> in such a way as to reveal to others that the actor can be trusted to assume the alignments and do the cognitive work required for the

> appropriate accomplishment of the collaborative tasks they are pursuing in concert with each others [*sic*], that is to act as a moral member of the community being sustained through the actions currently in progress. (2007: 70–71)

This treats morality as a very basic presupposition of ordinary interaction, the mere expectation that others are cooperative (see also Schegloff 2006: 85). We may not want to define the term so broadly (see Lempert 2013), but Goodwin's reference to trust points toward the normative force of default expectations against which alternative actions, or inaction, come to have significance. These may function as neutral devices merely to coordinate interaction. But as Goodwin suggests, they are available for ethical evaluations. Moreover, to the extent that these acts are done with regard to other persons and to the participants' shared commitment to the interaction, and to the shared reality that interaction sustains, it follows that their breach may also constitute an act of disregard for others. *Disregard*, in this context, refers to the ways in which people make sense of the flow of interaction itself. In Goffman's words,

> Evidence of social worth and of mutual evaluations will be conveyed by very minor things, and these things will be witnessed, as will the fact that they have been witnessed. An unguarded glance, a momentary change in tone of voice, an ecological position taken or not taken, can drench a talk with judgmental significance. (1967: 33)

Note the reflexive dimension: the fact that I witnessed something about you is itself something that you notice. Ethical life includes both this quality of reflexivity—in Sacks's words quoted earlier, Adam and Eve's sense of being observables—and therefore, also, the potential for ethical evaluations to take flight and circulate in public life well beyond the momentary encounter in the here and now. The techniques can serve everything from snobbery to sadism. Sidnell observes,

> One way to be rude is to specifically exploit the properties of the turn-taking system so as either to completely exclude some person from a conversation (disallowing them any opportunity to speak) or to withhold a response to what someone else has said and in this way to ignore them. (2010a: 37)

As we will see in the next chapter, much more than etiquette may be involved here. Such minor denials can be powerful assaults on someone's dignity and identity. They can draw on the propensity for discriminating among kinds of people that appears early in child development to target entire social categories. Such is the everyday life of racism and sexism. Exclusion can become systemic, a taken-for-granted ground for structures of political and social inequality. This is why social or religious reformers often make explicit and challenge interactive norms, as we will see in later chapters.

On a moment-to-moment basis, the shared reality that conversational interaction constructs and reaffirms—or denies—may be about the kind of activity participants are engaged in. But more can be at stake than just the logistics of traffic management. The ethical dimensions of shared reality hinge on the way that people are constantly making sense of one another, and to make sense of someone inevitably means to evaluate him or her in some way. In Goffman's (1967) depiction of "face-work," your face depends on others affirming that you are the kind of person *they* take you to be presenting to them. What is crucial is the role of the other persons. You cannot simply assert yourself; your self-presentation requires uptake by others on the basis of what it is they take you to be presenting. Agency is distributed across the social interaction in two respects, since you depend on others to recognize your "line" and their ability to do so depends on what they think you are doing. The vulnerability of the self is a function of this distribution. At stake is not necessarily just social status. It can be that element of self-knowledge that depends on seeing yourself from the outside. This is a consequence of the capacities for reciprocity of perspectives and for taking a third-person point of view on oneself that are among the critical cognitive developments discussed in chapter 1.

Because the construction of shared reality is a constant, ongoing process, and is inseparable from people's evaluations of one another, even quite minor effects are amplified. As Garfinkel's experiment with the students acting like boarders in their family homes makes clear, what is at stake is not merely conformity to some superficial, stylistic conventions. Minor interactions can have real consequences for how one is taken by others and how they take one's stance toward themselves. The proud men from El Barrio mentioned above found that they could not hold onto their jobs in midtown Manhattan office buildings. It is perhaps no accident that some of the more memorable moments of the American civil rights movement—Rosa Parks refusing to give up her seat on the bus, the Greensboro Four taking their seats at a drugstore lunch counter—focused on equally trivial-seeming interactions. We will see other examples of the serious consequences of apparently minor matters of interaction in chapters to follow.

If this is what character is—and there is no consensus about this—then many of the things that normative approaches might consider irrelevant to ethical life turn out to be significant. For example, as noted in chapter 1, both philosophers and psychologists tend to agree that people are innately prone to recognizing a distinction between universal ethical or moral principles, such as "Don't harm others," and local customs or expectations, such as "Only girls wear skirts." The usual claim is that even young children, when queried, will say that the latter may vary or change by fiat, but not the former. In this view, the latter is mere social convention.

The ethnographic evidence suggests two qualifications to this claim. The first, which I will return to below, is a matter of ontologies. For example, certain dress requirements, such as Islamic head covering, Jewish tefillin, or Mormon underclothing, may have divine mandate. To the secular philosopher they may seem matters of cultural convention, but not to the faithful. The second is the way in which clothing may be taken to reflect character. To dress "inappropriately" may not be just a matter of breaching custom; it may be taken to say something about what kind of person one is—for instance, as being uptight or of loose morals. Even where people recognize that certain norms are merely local conventions, to breach them might still be taken as evidence of failure of character. The failure may be a matter of expressing one's inner self or of manifesting respect or lack of respect for others. But why should one person's bad demeanor (for instance, foul language or revealing dress) offend someone else, if it is merely evidence of his or her own character? The evaluative dynamic is reflexive. I think that it is in this context that we can make sense of the parents' strong responses to their children's inexplicable politeness and distancing in Garfinkel's boarding house experiment. The breach of conventions was not merely a matter of stylistic choices. They were interpreted as evidence about the nature of relationships and evaluated in terms of ethical values, for instance, that members of a household should be on intimate terms because of who they are to one another.

Reflexivity imposes some constraints on this vulnerability of people to one another. The constraints derive from the very distribution of agency in play: I am not only concerned with my own face—most of the time I am also committed to yours. This is not merely due to altruism or empathy. *Deference*, the regard I display for you, is simultaneously a display of my own character as indicated in my *demeanor*, how I comport myself in the presence of others (Agha 2007; Errington 1988; Goffman 1967; Silverstein 2003a). That is, the practical dynamics by which one accords recognition to others is inseparable not just from their recognition of oneself but from one's own capacities for self-recognition. The self is ethically vulnerable in any situation in which it is sustained or defeated by the responses of others; but those others are not free of the implications of their own actions either.

The inherent vulnerabilities described here can, of course, be put to a strategic purpose. Deference and demeanor can function just as strongly to contest people's ethical standing as they do to affirm it. This has been nicely shown in the linguistic anthropologist Asif Agha's analysis of the debate between Bill Clinton and Bob Dole in the 1996 presidential campaign. The candidates faced a delicate challenge that required them to master the mutual entanglements of deference and demeanor. Each had to attack his opponent's character without damaging the viewers' perceptions of his own in the process of doing so. Agha's close analysis of the recordings of the debate shows the variety of ways each candidate took advantage of tensions among the explicit denotations of their words, body language, implicit allusions to previous utterances, and tacit invocations of the public's prior knowledge. Some of these effects are quite complex, but one of the more straightforward examples is the moment when Dole, referring to an old scandal associated with Clinton, says, "I've never discussed Whitewater . . . <gaze turns to Clinton>. . . . I'm not discussing Whitewater now" (Agha 1997: 488). At the level of denotation,

Dole is making use of the Ciceronian rhetorical device of preterition, to simultaneously say something and deny responsibility for saying it by stating just what it is you will not say. But the aggression is underlined by the body language, the meaningful glance toward the target. In this way Dole attempts to attack Clinton's character while avoiding the accusation that his own character is unduly aggressive. As Agha summarizes the debate, Dole's utterances tend to "convey a surplus of meaning beyond explicit denotation, and appear construable as 'aggressive' in some way" (1997: 480). Yet, because that aggression, while intuited by most viewers, could not be identified with the most easily reportable elements of speech—specific words and phrases—it was hard to pin on the speaker. As a result the next day the press described the debates as cordial and temperate (Agha 1997: 462). Dole's success seems to have been only partial (and of course he lost the election), but the reports in the press suggest that he managed to retain the surface forms associated with a courteous demeanor while repeatedly reminding the audience of prior attacks on his opponent's character.

The interplay between deference and demeanor is perhaps most evident in situations like the Dole–Clinton debate, where the participants are primarily interacting for an audience of outsiders to the interaction and their formal equality in the debate format makes them more or less equally vulnerable to one another. Such parity is, of course, not always the case. Nor is individual agency always as effective. Consider an imputation of character within the strongly inegalitarian conditions of colonialism and racial stereotyping, as analyzed by the linguistic anthropologist Judith Irvine (2009). In 1868, in the British religious mission in Onitsha, in present-day Nigeria, a violent brawl had broken out among the missionaries. Eventually the London office expelled the head of the mission, J. C. Taylor. A letter from the bishop of Lagos to London had complained of Taylor's character:

> I myself did suspect him to have been a little overpowered twice by the influence of ardent spirits, through his *unusual loquacity*, though it did not amount to actual drunkenness. . . . When he went on board these ships at night in October last, and wrote me such *disconnected notes, breathing hostile feelings* against all the mission agents, . . . his last

> desperate acts [are] very much suspected to be excited by the same influence. . . . This habit, combined with a haughty and irritable disposition, must disqualify him for his spiritual duties among the heathens. . . . You will perceive that I did not send you copies of Mr. Taylor's [earlier] letters . . . because they appeared to me *so unconnected, written with excited mind* and in other respects *very personal.* (Irvine 2009: 61)

What is striking here is the weight the bishop gives to what Irvine (following Goffman) calls "faultable aspects" of Taylor's linguistic performances, shown in italics. Here are some examples of those faultables in Taylor's letters to the bishop:

> You have much to think & do than to listen to unpleasant news.
>
> I overheard Mr. Coles was bawling out loudly in the yard, having heard the person who brought the slave when he was about going said quietly.
>
> Then came in Mr. Phillips & the women I have named, my revolver was on the table as we are troubled with leopards. I held it in hand & told them to walk out of my parlour, then Mr. Cole jumped upon me at once, & Mr. Phillips pretending to take Mr. Cole away whilst his wife biting me & the beatings I have received from them. (Irvine 2009: 62)

Irvine points out that there may have been several reasons for these dysfluencies, including extreme emotional distress and severe physical pain from the attack. Moreover, Mr. Taylor was not a native speaker of English, his mother tongue being Igbo; similar dysfluencies occur in letters written by other Africans involved in this mission. But the case against Taylor had, in effect, already been settled. The meaning of his words and acts was not his alone to determine:

> African missionaries always walked a thin line. They were always vulnerable to accusations, either that they were arrogant (uppity, so to speak) or that they had slipped "back" into venality, carnality, and disorder. . . . In these circumstances, it is the observers of the performance who make the attribution, who find a way to map the stereotype onto the performance no matter what. (Irvine 2009: 68–69)

As this episode shows, even apparently minor aspects of demeanor, such as grammatical errors, can contribute to the public construal of a person's character. Moreover, the construal of demeanor is never entirely in the hands of the individual, since it depends on uptake by others. In this respect, the character that emerges is co-constructed by the individual and others. That co-construction is likely to draw on what others think they already know—about the individual and his or her *type* of person (African, for example). For example, the linguist Susan Ehrlich points out that male defendants in rape trials—a particularly consequential forum for giving an account of oneself—often rely on the prevailing assumption that men and women are naturally prone to misunderstand one another in order to claim their innocence. Summarizing her analysis of one case, she writes, "What Matt [the defendant] and the tribunal characterized as 'miscommunication' is better understood as culturally-sanctioned innocence" (2001: 148).

As Haviland's analysis of gossip in Zinacantan suggests, imputations of character, while not immutable, are beyond the capacities of any single person to define. For this reason, character also remains vulnerable to an unpredictable future. First, the focus is on character, not acts, and therefore it extends over time beyond the limits of any specific action or event (with variations: anthropologists [e.g., Robbins and Rumsey 2008] observe that not all communities expect individuals to display consistency of character over time). Moreover it draws on inferences about the past (one was once a cheater) and possibly the future (and therefore will always be a cheater). Those inferences invoke aspects of the self that are not present in the here and now of the immediate interaction. But because the establishing of character is interactive—it depends on other persons—it can never be definitively settled once and for all: it remains in play as long, presumably, as there are others to interact with.

Finally, the co-construction even of bad character is something to which all may be complicit, even those who bear the brunt of it. Referring to the internalization of positions within the class hierarchies of eighteenth- and nineteenth-century England, the political theorist Don Herzog asks how anyone could be complicit in her own social subordination:

> It's not just that she is subjected to a deafening roar constantly instructing her that she is inferior, and willy-nilly comes to adopt that belief, as though a message bombarded frequently enough finally imprints itself on the target. It's that she finds herself in a role, with a job to do. In doing it, in digging in and negotiating the daily demands of her work, she develops certain character traits, works up an account of the world she lives in, gains characteristic virtues and vices. She exercises agency. And she may sculpt herself into a happy slave, cheerfully contemptible. (1998: 361–62)

The failure to internalize dominant norms, of course, is also well attested, as are its consequences. Describing how tough drug dealers from New York's El Barrio fail in the context of middle-class life, Bourgois observes:

> Street culture is in direct contradiction to the humble, obedient modes of subservient social interaction that are essential for upward mobility in high-rise office jobs. . . . [Dealers] do not know how to look at their fellow service workers—let alone their supervisors—without intimidating them. They cannot walk down the hallway to the water fountain without unconsciously swaying their shoulders aggressively as if patrolling their home turf. (2003: 142–43)

As with Herzog's "happy slave" or Bourdieu's Kabyle woman mentioned earlier, values are not simply some external system of concepts or rules but, rather, a way of being a person in a body and being aware that this makes you visible to others—that one is "an observable."

To be sure, there are important distinctions among these three examples. The working-class woman of whom Herzog writes has internalized and, it would seem, even rationalized the contempt in which others hold her. The Kabyle woman, perhaps, has internalized an external view of herself that we should not assume to be contemptible—until, perhaps, she finds herself in France. Like her, the men of El Barrio also discovered their embodied self-esteem to become a mark of stigma when the context changed and the people observing them shifted from neighbors uptown to office workers in midtown. So their character is not merely a matter of reputation: bodily habits may not go all the way down, but they go deep. On the other hand, the ethical values embodied in those habits are inseparable from their concreteness, their availability to the judgments of others. The body is one basis for monitoring the self and its failures, as with the person whom Warner invokes, who finds himself speaking too loudly in a quiet room, or the El Barrio drug dealer whose swagger comes off as aggression. In either case, these are not simply wholly autonomous selves that happen to be staging themselves for others.

Let me summarize the argument of this chapter. Ordinary conversation is governed by devices for regulating smooth-flowing interaction and constructing a shared sense of reality (Atkinson and Heritage 1984; Sacks, Schegloff, and Jefferson 1974). But these devices are not merely technical means of traffic control and consensus establishment. Independent of technical functions, they offer affordances that people can draw on as they evaluate one another. Central to this evaluative dynamic is the ongoing work of projecting a sense of self and responding to the sense of self projected by others. Evaluation draws on and reinforces known and presupposable types and categories of person and action, which are themselves potentially evaluative. There is an ethics of commitment both to the shared reality and to the sense of self of the respective participants. Of course that ethics is constantly challenged. But those challenges depend on the ways in which people are emotionally attached to their own "face"—that is, to how they see themselves through the eyes of others—and the vulnerability that ensues from the dependence of face on being affirmed by others. Thus commitment to the conversation easily becomes commitment to the other person. Because the technical details of the flow of conversation themselves have implications for the regard in which people hold one another, and the degree to which they do or do not share a sense of reality, of what is actually going on in any given moment, all of this is mediated by everyday semiotics. In the next two chapters we will see some of the ways these habitual and unself-conscious aspects of interaction have been taken up as objects of reflection in different cultural circumstances. Attending to the processes that induce reflection can give us greater insight into ethical life as something that involves self-conscious people who live in particular communities and draw on historically specific practices, ideas, and ways of talking.

CHAPTER 3

Problematizing Interaction

Dignity and Respect

One objection to placing a heavy emphasis on the semiotics of interaction is that this seems to reduce the whole business of ethical life to mere appearances. The philosopher Alasdair MacIntyre, for instance, accused Erving Goffman of harboring a cynical world view, in which "imputations of merit are themselves part of the contrived social reality whose function is to aid or to contain some striving role-playing will" (MacIntyre 2007: 116). But the vulnerability of the self is not just a matter of stage management and strategic manipulations that seek to align appearance with a fixed set of external social standards. I have tried to show that interaction is crucial to the constitution of ethical sensibilities (as well, one might surmise, as important aspects of "the will") in the first place. To get at the stakes, I would like to turn briefly to the philosophical idea of "dignity." The purpose in doing so is not to impose a particular philosophical system on a highly variable ethnographic record. Rather, the idea of *dignity* in English can help make visible the ethical dimensions of ideas found in other ethnographic contexts. This does not mean that all such ideas are simply versions of dignity, only that parsing them together can draw out certain themes they share and clarify how they differ. Dignity and the other concepts discussed in this chapter are all specific ways in which different communities have reflected on and responded to potentially ethical dimensions of cognition, the emotions, and social interaction.

One of the most famous invocations of dignity in Western thought is Kant's distinction between ends and means:

> In the kingdom of ends everything has either a *price* or a *dignity*. What has a price can be replaced with something else as its *equivalent*;

> what . . . is raised above all price and therefore admits of no equivalent has a dignity. . . .
>
> . . . [T]hat which constitutes the condition under which something can be an end in itself has not merely a relative value, that is, a price, but an inner value, that is *dignity*.
>
> Now, morality is the condition under which alone a rational being can be an end in itself, since only through this is it possible to be a lawgiving member in the kingdom of ends. Hence morality, and humanity insofar as it is capable of morality, is that which alone has dignity. (1996: 84)

This expresses one of the sources, for Western thought, of the idea that human ethical life must be treated as something that, by its very essence, is a distinct kind of entity from the objects of naturalistic explanation. Here we will consider the role of respect and demeanor in the history of the idea of dignity. For it is in these registers that abstract principles can appear in the worlds described by ethnography. We can see distant echoes of the Kantian idea even in the demotic and seemingly amoral world of crack dealers in El Barrio:

> Primo's best and cheapest insurance against physical assault was to surround himself with a network of people who genuinely respected and liked him. [But when I mentioned this,] Primo considered somewhat insulting my functionalist interpretation of why he treated his friends and acquaintances so generously. (Bourgois 2003: 107)

Even someone as cynical as a drug dealer, in finding the suggestion that he used his friends offensive, was nodding to the premise that at least some humans are not supposed to be treated only as instrumental means to other ends.

The philosopher Michael Rosen summarizes the Kantian position this way: "The dignity of the moral law makes human beings—its embodiment—worthy of respect. They should be respected by others and . . . they have the duty to respect themselves" (2012: 30). Notice the reflexive quality of this: one should respect oneself. Like the interaction of deference and demeanor noted in the previous chapter, here the respect one owes to others mirrors the respect one owes to oneself—and vice versa. In the Western tradition, Rosen points out, the idea of dignity originally applied to people of high rank and slowly expanded in scope, eventually encompassing humans as such. We will return to the historical expansion of the range of moral concern in later chapters. What I want to stress here is the role of demeanor. Rosen writes,

> To treat someone *with dignity* is . . . to *respect* their dignity. . . . To respect someone's dignity by treating them with dignity requires that one *shows* them respect, either positively, by acting toward them in a way that gives expression to one's respect, or at least, negatively, by refraining from behavior that would show disrespect. (2012: 57)

It is this sense that the moral philosopher Stephen Darwall (2006) develops in his idea of the "second-person standpoint." Darwall argues that the importance of dignity as a moral concept is not just in the first-person sense that "I possess dignity." Rather, what is crucial is that I can demand that another person treat me with respect for my dignity. For Darwall, this is the basis for human rights. Although Rosen has a somewhat different approach, he too takes the treatment of others with dignity to be foundational for rights. What I want to stress here is the fundamentally *interactive* character of the idea of dignity and the moral weight that both philosophers accord it.

Notice, however, that dignity can draw on the third-person perspective of a larger public world beyond the parameters of any immediate interaction. We can see this in the example of a 1995 French court case Rosen discusses. After a town prohibited the game of dwarf-throwing in a local pub, a dwarf sued, claiming that the ordinance violated his right to employment. He lost the suit on the grounds that dwarf-throwing, by offending the dignity of dwarves in general, was a matter of public morality (Rosen 2012: 63–65). But how one treats others reflects not just on them but on oneself. Rosen points to how the living handle the dead: "We have a duty to treat a corpse with dignity just because one of the ways in which we have a duty to act is that we should perform acts that are expressive of our respect" (2012: 139–40). Even if the dead cannot benefit from the way we handle them, it still matters for the self-respect of the living. Recall that deference and demeanor are interlocked: in a similar way, by treating the dead properly, I display my respect for them and, in the process, enact my own character as the kind of person who respects the dead. Here the concept of dignity has been removed entirely from ordinary interaction, yet retains something "exterior"—that is, a form of material activity that mediates the relations between self and other—that is a core component of ethical life. Dignity in these accounts is an ethical quality that cannot be understood just in reference to the individual to whom it pertains: there must be someone else who respects it. Second, it follows from this that the ethics of dignity depends on a semiotics of behavior. One person's dignity must be perceptible to another; it cannot remain an inner quality alone. Or, if we follow Darwall, one person's demand that another person respect his or her dignity must take palpable form.

Ethnographies, unsurprisingly, are full of descriptions of the resources and strategies that allow people to retain their dignity, even in the face of apparent failures of face. For example, Bourgois notes that New York crack dealers who had lost their low-level office jobs were likely to see this as resulting from their principled refusal to abandon their dignity. They treated economic failure as a form of self-assertion in the name of deeper values:

> Obedience to the norms of high-rise, office-corridor culture is in direct contradiction to street culture's definitions of personal dignity—especially for males who are socialized not to accept public subordination. . . . They were usually fired from these jobs, but they treated their return to the world of street dealing as a triumph of free will and resistance on their part. (Bourgois 2003: 115)

As one of these men reflected, after being fired:

> I got my respect back. . . . Buela [grandma] likes me. I got a woman. I got a kid. I feel complete now. I don't really need nothing. I got money to get wrecked. . . . The money I make in the Game Room [dealing drugs] is for my personal madness; for my personal drug-addiction and self-destruction. It's something only I could control. No one could tell me what to do with it. (Bourgois 2003: 118)

He had his dignity back. We might say that he was reasserting the ethics of mutual recognition: his grandmother likes him, a form of dependence that seems inseparable from his assertive independence, needing nothing, having something he can control. But his recognizability to himself and to others depended on selecting those with whom he would interact. As ethnographers and sociologists of class have long recognized, when this self-preservation depends on remaining in the context of the family, neighborhood, or class within which embodied norms are well known, there may be a steep economic cost in curtailed opportunities.

For men of El Barrio, dignity is associated with the aggressive display of masculinity. But we can also see the ethics of dignity at work in a quite different world, that of militantly gay men:

> In those circles where queerness has been most cultivated, the ground rule is that one doesn't pretend to be *above* the indignity of sex. And although this usually isn't announced as an ethical vision, that's what it perversely is. In queer circles, you are likely to be teased and abused until you grasp the idea. . . . This kind of culture is often denounced as relativist, self-indulgent, or merely libertine. In fact, it has its own norms, its own way of keeping people in line. I call its way of life an ethic not only because it is understood as a better kind of self-relation, but because it is the premise of the special kind of sociability that holds queer culture together. (Warner 1999: 35)

Notice that the assault on dignity Warner describes here depends on shaming practices—it still takes place within the domain of self-esteem and its dependence on the respect of others, a domain where we also find dignity.

If deference, demeanor, dignity, and respect are to serve as analytic tools, they should help us understand ethical life beyond the West, for it may be that even crack dealers and radical queers still take their bearings from the background assumptions of modern Euro-American societies. So let me turn in somewhat more detail to the concept of *dewa* as I understood it while living on Sumba in the 1980s and 1990s. Several decades earlier, the missionary-ethnographer Louis Onvlee summarized the concept in ways that still rang true in my time. *Ndewa* (as he spells it) is "that in a man through which he is as he is, . . . his character and to an important degree, also his appointed fate, and that which differentiates one person from others" (1973: 211–12). Notice the connection between fate and individual identity. If fate is what distinguishes random events from a biography that has a direction (Fortes 1959), then it may, for example, be heading toward success or failure. Dewa links that narrative to individual distinctiveness, but not one that is created by the individual alone.

Living in Sumba, I heard the word *dewa* used most often when people talked about their interactions with others. In particular, it explains why one does or does not have influence over them. This influence is crucial in a society where a highly elaborate system of ceremonial gift exchange in marriages, funerals, and other rituals is both unenforceable and yet central to the forging of social relations (Keane 1997). The exchanges that make social relations tend to have a competitive tone to them. Even though marriage negotiations should end with a solid alliance between families, as noted in the introduction, they are often portrayed as a contest. The successful suitor might say that his dewa defeated that of his bride's father. When gift exchanges do not go well, it may be due to a disappointed recipient having a "hindered dewa" (*patipang dewa*), which rejects other people's generosity. If other people *are* generous to you, that shows the power of your own dewa. This is why you should never turn down even a trivial gift, such as a chew of betel nut or a cigarette; to reject good fortune could block future generosity. This is also one reason petty traders who try to operate where they have many relatives often fail. The ethics of interaction overwhelm the dictates of profit-seeking. As one of them told me, "If you give people credit, they'll just feel honored, and won't feel any need to repay the debt" (Keane 1997: 203). In this way, the idea of dewa both acknowledges how dependent people are on one another and yet also ends up assimilating other people's actions to one's own character.

The dewa is manifested not just in the outcomes of interaction but in a person's bearing. Like the virtue of the Kabyle woman described by Bourdieu, or the masculine self-esteem of the crack dealer portrayed by Bourgois, the condition of one's dewa is embodied and, therefore, is perceptible: it is semiotically available to others. At the same time, it palpably registers the effects of those others on oneself. The result of being offended by someone else is a "small dewa" (*meduku dewa*), which shows that someone doesn't know his or her "own value" (*wili wikina*). Here's how one man described it to me:

> When we gather, such a person walks timidly, doesn't speak up, like if he comes to our house, he doesn't directly climb up on the veranda where we sit and stick out his hand, but creeps along to the far end and sits there. . . . [I]f he makes a conveyance oration, his enunciation isn't right. Or his clothing isn't suitable. (Keane 1997: 206)

In its stronger forms, a small dewa can lead to absentmindedness, fainting spells, or shamelessness, in short, a loss of self-possession. But, like dignity, dewa is not just a matter of individual character. A weak or hindered dewa may come from being embarrassed, startled, or insulted. That is, it is the result of interaction gone wrong, which in turn threatens to induce further bad interactions in the future. Sometimes a hindered dewa results from a mismatch between husband and wife. In these ways, the concept of dewa seems (in distinctively Sumbanese terms, which include the agency of spirits and the magical power of material things) to thematize the role that other people play in one's *own* sense of self.

This concept, however, is not simply just one way of describing a universal feature of interaction, a local translation of a word we might also find in vernacular English or the technical vocabularies of philosophy and psychology. Once crystallized as an object of reflection, something that Sumbanese consciously know about the world and can connect to other things they know about their world, it also guides them as they purposefully undertake ethical actions. Here is just one example. A man I call Ubu Kabàlu had a nasty encounter with a government cattle inspector, who roughed him up, using harsh language. In order to "make the spirit return," Ubu Kabàlu sponsored a big feast for all his relations and allies. Feasts normally involve not just hospitality on the part of the host but gifts from the visitors, such as horses, pigs, valuable textiles, and gold or metal ornaments. So by compelling others to flamboyantly display their respect for him—perhaps something like Darwall's demand for dignity—Ubu Kabàlu was able to restore his dewa from the perilous condition into which it had fallen. Otherwise his weakened dewa might well have damaged his social interactions into the future. Even though the guests involved in the feast were not identical to the man responsible for the original insult, their recognition had effects on the wounded dewa. For instance, after the feast Ubu Kabàlu could hold himself tall and expect his dewa to exert greater power over his exchange partners. To summarize, the idea of dewa takes aspects of interaction like those Goffman describes in terms of "face" as affordances that, transformed into a knowable object, are part of a constellation of other concepts such as the agency of spirits, which helps give a distinctive shape to ethical self-awareness and action.

VARIATIONS ON INTERSUBJECTIVITY

One of the distinctive features of the concept of dewa is the way it allocates intentionality. More strongly than dignity, dewa posits that ethical effects do not need intentional actors. On the one hand, it portrays the giver's generosity as due to the power of the recipient's dewa. On the other hand, it also shows that one person's failings register the effects of others on his or her dewa. As suggested by P. F. Strawson's description of resentment quoted in chapter 2, this alertness to intentionality indicates that dewa has, at its heart, a fundamentally ethical concern. How you ascribe intentionality affects whether an action is considered ethically consequential at all (rather than, say, accidental or neutral), who bears responsibility for it, and to whom it is relevant. The kinds of situations that tend to evoke talk about dewa are ethically fraught—the making or breaking of social relations and the self-esteem or imputations of character that go with them. But they can also involve merely the loss of composure, given the right circumstances. This is what happened when one elderly lady stumbled while climbing onto her host's veranda during a funeral. This seemingly trivial misstep was taken by everyone to threaten the stability of her dewa, prompting her host to leap up and present her with a fine textile to restore her self-possession.

Like any theory, philosophical accounts of intentionality are products of specific traditions. The twentieth-century English-speaking world within which Strawson writes about intentionality and resentment is notable for its emphasis on the autonomy of individuals and the importance of personal intentionality, understood to be found inside the self, for understanding one another's actions. When faced with such views, anthropologists have tended to be skeptical of the idea that individual intentionality plays the same role in all social worlds or, indeed, any role at all.

In a moment, we will consider an especially strong version of this anthropological skepticism about universals of intentionality. But first, let us take a glance at two more examples. Here I draw on studies, written in a classic ethnographic mode, that portray distinctly non-Western ethical worlds in which ideas about intentionality, or something close to it, are prominent. The Chewong of highland Malaysia and the Inuit of Arctic Canada are both societies of hunters or hunter-gatherers (the present tense here refers to the time of the fieldwork in question, the late 1970s and early 1960s, respectively), who live in very small, ideologically egalitarian communities. Sumbanese dewa reflects life in a society where gift exchange mediates hierarchical relations between people who identify with larger social units and see their lives as guided by ancestral rules (Keane 1997). By contrast, Chewong and Inuit are highly mobile, and their social relations have a strongly voluntaristic quality. Individuals have a great deal of choice about whom they associate with and easily leave one settlement for another. People in these communities seem to have developed an ethical language that crystallizes some distinctive stances toward the purposes and desires of those around them. Here, for example, is the concept of *punén* reported of the Chewong. *Punén* refers to a condition that results in attacks by dangerous animals such as tigers, snakes, or poisonous millipedes or diseases caused by their spirits. The condition is triggered by the victim's suppressed desire. One type

> has social ramifications since it prescribes the sharing of food and other objects, and prohibits the nursing of ungratified desires. . . . If someone is not immediately invited to partake of a meal which he observes, or if someone is not given her share of any foodstuff seen to be brought back from the jungle, that person is placed in a state of *punén* because it is assumed one would always wish to be given a share and hence experience an unfulfilled desire. (Howell 1989: 184)

Like the Sumbanese concept of dewa, punén seems to focus on the reflexive and intersubjective nature of social interaction. It is based on a fundamental presupposition, that my possessions prompt desire in others:

> The Chewong take all possible precautions against provoking *punén*. All food caught in the forest is brought back and publicly revealed immediately. It is then shared out equally among all the households. . . . If guests arrive while the hosts are in the middle of a meal, they are immediately asked to partake. If they refuse, saying they have just eaten, they are touched with a finger dipped in the food, while the person touching says "*punén*." (Howell 1989: 185)

Notice how the consequences of *my* refusal to share rebound not on me but on the *other* person. The ethics of interaction therefore focuses on protecting others against the consequences of their own desires, understood in interactional terms. It is the responsibility of the owner of the food or object to ward off the danger that desires for it pose to the other person—something that is provoked in the moment of social encounter. It is striking that, when not actually sharing the thing in question, the prophylactic is itself reflexive, explicitly naming the condition that is to be avoided. This raises the question of ethical categories and explicit talk about them, to which we will return in the next chapter.

Another way of focusing on the ethics of interaction in a small-scale, egalitarian world is that of the Inuit (known at the time the ethnography quoted here was written as the Eskimo):

> The feeling Inuttiaq was expressing is one that is very characteristic of Eskimos: a fear of people who do not openly express their good-will by happy (*quvia*) behavior, by smiling, laughing, and joking. Unhappiness is often equated with hostility (*ningaq*, *urulu*) in the Eskimo view. A moody person may be planning to knife you in the back when you are out fishing with him, claiming on return that you drowned. In the old days he might have been plotting to abscond with your wife. . . . Even without harboring specific evil designs an unhappy person may harm one, merely by the power of his moody thoughts. It is believed that strong thoughts (*ihumaquqtuuq*) can kill or cause illness;

> and people take great pains to satisfy others' wishes so that resentment will not accumulate in the mind. A happy person, on the other hand, is a safe person. (Briggs 1970: 47–48)

Descriptions like this are usually treated as a form of ethnopsychology. What I want to draw attention to here is their ethical dimension. First, they have direct implications for local ideas about character and the self-cultivation that goes into it. As the anthropologist Jean Briggs writes, "Emotional control is highly valued among Eskimos; indeed, the maintenance of equanimity under trying circumstances is *the* essential sign of maturity, of adulthood" (1970: 4). In effect, here equanimity is a kind of virtue. But self-control is also a response to the danger posed by others. Moreover, this danger that people present to one another underwrites a distinctive ethical semiotics. It leads to the ethical injunction to behave as if they were no threat at all, by performing goodwill.

In Briggs's treatment, however, attentiveness to the forms interaction takes is a direct expression of people's sensitivity to the perspective of others and the consequences for their own self-esteem:

> Although the ideal says that "everybody" should be helped, in reality, help is extended much more willingly to close relatives than to others, and people, knowing this, sometimes prefer to do without, rather than appeal to outsiders, if what they need cannot be provided by close kinsmen. The reluctance to ask from people outside the extended family is phrased as a fear of being unkindly refused. . . . Similarly, a major motivation for giving to people in the larger group is the fear of being thought unkind. . . . So great is the embarrassment of refusing and being refused that requests are, as a rule, made most indirectly; if one wishes to ignore the hint one can do so. . . . Very frequently, moreover, the responsibility for requests is attributed to someone else. One may say, "So-and-so told me to ask" (assuming a cloak of docility—a Good quality), or: "I ask because so-and-so is cold" or "hungry" (presenting oneself as generous and thoughtful). (1970: 209–10)

Chewong and Intuit, in these descriptions, seem to inhabit intimate worlds in which, it would seem, there is a great deal of consensus about what people are like and what they can expect of one another.

In both cases, people seem to have no hesitation about imputing desires and intentions to others on very slim evidence. Theirs is essentially a dispositional approach to interaction. The presupposed dispositions become a default against which action or inaction can be evaluated. As Briggs remarks of the Inuit,

> It sometimes happens that even when there has been no request [for assistance], a failure to *offer* assistance spontaneously will be interpreted as deliberate unkindness. And when people, whoever they may be, are slow to offer help, or show themselves jealous and greedy . . . , the wider values, "people should be good to everybody," are invoked against them—not, by and large, in direct accusations, but in private gossip or in the moral generalities of Sunday sermons. (1970: 210)

Like any ethnography, these accounts have their limits. Written in a fairly traditional ethnographic style, they aim to characterize a collective ethos in a relatively timeless framework, treated without much reference to the "external" sociopolitical context, and rarely enter into details of interaction. But granting those limits, they are useful for showing how intersubjectivity and intentionality have been thematized in quite distinct sociohistorical contexts.

Contrary to one tradition in ethnographic writing, we need not take the Chewong or Inuit to be creating their assumptions and ethical judgments on unique and ontologically self-contained grounds. They are not entirely unfamiliar. Briggs's ethnography became famous because it pioneered the foregrounding of the ethnographer's own emotional experiences and drew unsparing portraits of distinct Inuit individuals. She opens the book with a story about losing her temper at some nonindigenous Canadians who had damaged an Inuit canoe. To her hurt and bewilderment, the Inuit responded by ostracizing her for three months. Much of the book is an exposition of the Inuit's assumptions that underlay this surprising reaction, centering on the danger they see people's emotions pose to one another. Like Bourgois's account of crack dealers, this ethnographic procedure works by taking the apparently strange and perplexing and persuading the reader that he or she can, in the end, understand its internal emotional and conceptual logic. If this procedure is not entirely erroneous, then, it depends on the reader finding something recognizable—perhaps people's ethical vulnerability to one another—in what had at first seemed so alien.

UNDERDETERMINED EMOTIONS, SPECIFIC CONCEPTS

The concepts crystallized in words such as the English *dignity*, the Sumbanese *dewa*, El Barrio's *respect*, the Chewong *punén*, and the Inuit *ihumaquqtuuq* pose, of course, significant problems of translation. Each is embedded within the vocabulary and grammar of a particular language and linked to a host of other ideas and practices. We should not assume they denote distinct entities that exist independently of the larger conceptual and linguistic contexts in which they function. This is a long-standing topic in debates about epistemology (Davidson 1984; Quine 1969; Silverstein 2003b). Here, however, I want to show how these words also stand for a variety of ways in which people in particular sociohistorical circumstances respond to experiences that have sources in ubiquitous cognitive, emotional, and interactional patterns that may run across conceptual and linguistic contexts. This does not mean that concepts such as dignity or dewa are simply different names for the same things. Rather, it seems that they direct attention and give specificity to aspects of experiences based on objective sources whose meanings are underdetermined. In doing so, they also link them to other experiences and ideas that are also salient within a given community: for instance, whereas dignity might take some of its conceptual shape from its ties to ideas about the value of individual autonomy, dewa points toward people's mutual dependence. But given any particular set of concepts, developed within a certain social context, those cognitive, emotional, and interactive sources will harbor other possibilities that in the instance remain more or less unacknowledged. Those very possibilities may, however, be what some other community fastens on and elaborates. Again, the basic cognitive, emotional, or interactive phenomena are not simply things waiting for someone to name them. But neither are dignity, dewa, and the rest simply cultural inventions created from scratch. They are outcomes of processes of objectification that result in particular ways of responding to the affordances that cognition, the emotions, and interaction offer. Since the sources of specific ethical concepts are underdetermined for meaning and the other concepts and practices with which they articulate enter the given social context by other historical trajectories, the consequences cannot be predicted from the nature of their "raw materials."

To grasp what I mean by *underdetermined*, consider the feeling of anger. As we saw in chapter 1, whether a given state of neurophysiological arousal does or does not result in you feeling angry may depend on contextual factors, such as what information is available to you or how the people around you are acting (Larsen et al. 2008; Schachter and Singer 1962). That state of arousal, or something similar to it, thus seems to be a necessary precondition for the emotion of anger, but it is not sufficient to determine that "anger" is what results. As also noted in that chapter, Gibbard (1990) proposes that a feeling such as anger is only a moral emotion if it is about something that it makes sense to feel angry about. That quality of being "about" something is not inherent in the neurophysiological state that might come to be called "anger." Anger viewed as a moral emotion is thus part of a particular vocabulary of ethical concepts, such as *dignity*, *dewa*, *punén*, or *ihumaquqtuuq*. Each of these terms takes its meaning from its place within a larger constellation of ethical and psychological concepts, practices, and institutions. This is one reason why we should be cautious about the use of English-language words such as *anger* in research on emotions, since they may unwittingly smuggle the categories of a local ethnopsychology into findings meant to apply to all humans. These are the contexts within which something does or does not make sense to have a particular moral emotion about. For example, what it makes sense to feel angry about will vary from one social or historical context to another. You can only be angry at an offense to your honor in a social world where honor really matters. Or the degree of anger will vary with the local salience of the ethical category. In one famous experiment, working with the premise that the cultural concept of honor is more elaborated in the American South than in the North, researchers found that Southern males displayed a physiological response to offenses against their honor far exceeding that of men from the North (Cohen et al. 1996). As old targets of justified anger disappear, new ones emerge. In the latter case, you may at first feel anger without being able to point to a source, only later discovering what that feeling was about and why it mattered. This experience of shaping a formerly inchoate emotion was important for the feminist consciousness-raising groups we will meet in chapter 5. Their contribution to ethical history was in part due to defining and legitimating new things it made sense to be angry about, such as "sexism." All this suggests that identifying one's feelings of arousal as having a particular moral object such as righteous indignation or feelings of betrayal will have effects on the resulting emotional experience that cannot be discovered directly in the neurophysiological sources of those feelings.

It would seem that a given conceptual and linguistic context constructs ethical worlds whose particularity lies in their internal relations, not in the more or less raw, possibly universal, materials with which they work. But that particularity does not depend on those ethical worlds being entirely self-contained and self-constructing cultures. The ethical worlds that give rise to concepts such as dignity, dewa, and the like are not utterly closed off to one another. One reason is that they draw on aspects of cognition, affect, and interaction that humans share by virtue of their common natural history. Another, however, is that human social worlds interact with one another by virtue of their social history. Indeed, as I will argue in the next chapter, the coexistence, clashes, and convergences among historically constituted ethical worlds are crucial stimuli to moral reflection. This ability to clash and converge, however, depends on some degree of recognizability—even if it turns out to be misrecognition. Before elaborating these points, however, let us consider one strong counterclaim, the denial of intention-seeking or mind reading reported of some societies.

THE OPACITY OF OTHER MINDS

In the previous chapter I invoked Strawson and Grice, both of whom put people's tendency to read intentions at the heart of the ethics of interaction. The obvious question arises as to whether the orientation to other minds sketched here holds beyond the Euro-American contexts in which it has usually been described. After all, Strawson and Grice both speak to us from particular societies at particular historical moments, drawing on the words and ideas of traditions that place a heavy emphasis on individual autonomy, inner thought, and some version of free will. A strong counterclaim is that there are other societies (for instance, in Melanesia and elsewhere in the Pacific) that maintain an "opacity doctrine," the belief that "it is impossible or at least extremely difficult to know what other people think or feel" (Robbins and Rumsey 2008: 407–8). This doctrine is one example of how ethnographic evidence seems to push the range of human possibilities beyond the universals reported by psychologists (for an important reappraisal see Duranti 2015: 2–15). Here I will follow the more recent ethnographers who work with people who make opacity claims to propose an alternative perspective on them, one that seems to be illuminated by the idea of affordance.

The opacity doctrine, the claim that we cannot know what is in the mind of another person, has sometimes been treated as evidence of deep, culturally specific, cognitive differences between people in certain societies who speak this way and those in other societies who do not. But recent research suggests that the doctrine is about something more limited, the relations between public evidence—how people talk and act—and private states—what they think, feel, intend, or desire. For example, it turns out that in many cases Melanesians and Polynesians who do not meet adult norms, such as children and people who are mentally ill, openly impute intentions to other people (Besnier 1993; Schieffelin 1990, 2008). In other words, opacity claims are not about whether people have *access* to others' intentions but, rather, whether one should *talk about* those intentions. At the heart of the opacity doctrine, then, is the problem of hiding one's intentions from others. When people in Melanesia talk about this problem, they may evoke certain images of the person's hidden interior. One of them is what we could call the "inner theater," in which the self is divided into a speaker and an addressee. For instance, the Korowai of West Papua speak of their thoughts in quotation frames (Stasch 2008: 444–45). They depict thought as introjected social interaction. The heart is the speaker, and the reporting "I," the person who reports on the words of the speaker, is the person who carries out the action in response to the heart's words. The second image for that hidden interior is the pocket. In many Melanesian societies, your pocket is where you can keep goods out of sight of those who might make a claim on them. This forges a potent link between the material exchanges that mediate social relations and the hiding or revealing of one's inner thoughts. The image depicts relations between thought and word as parallel to other ways people forge or deny relations with one another such as gift-giving.

INTERIORITY

In communities where the opacity of other minds is stressed, interiority and intentions come into especially sharp focus when people convert to Protestant Christianity, where norms of sincerity are often highly elaborated. When the Urapmin of Papua New Guinea converted, for instance, their views on people's opaqueness to one another stood in direct contradiction to teachings about the Christian conscience (Robbins 2004). The problem of interiority was thematized by the religious practice of confession. Confession demands sincerity, or at least, words and thoughts should match up (Keane 2007). In this context, the display of one's inner thoughts to others is a reply to the demand by someone else, the priest, that one put one's thoughts into words for him. Faced with this unprecedented practice, Urapmin at first began confessing to the sins of *other* individuals. Why? It seems that they were responding to the insight that once thoughts can openly become words, they can enter into public circulation, liberating one person to say what another person is thinking.

The anthropologist Joel Robbins reports that the Urapmin found it easier to learn how to *speak* confessions than they did to *hear* them. It seems they found that to hear another's confession is to essentially be put on the spot and embarrassed. Confession in Melanesia often circulates around ideas of shame, the exposure of one's loss of the ability to keep something hidden. The focus is thus not on the knowability of inner thoughts, as an epistemological problem, but on the capacity to hide them, as a practical, ethical, and political problem. So why should it be shameful to hear people talk about their inner thoughts? In many of the cases here, it seems to be because you are not just intruding on other people's interior or private space; you are being made witness to the embarrassment of seeing them lose that ability to keep hidden what they ought to have kept hidden. The problem is not only psychological or epistemological. The problem concerns a person's capacity to hide his or her inner thoughts from others. It is not that inner thoughts are inherently unknowable but that they ought to be unspeakable, or at least, it matters greatly who gets to speak those thoughts—that is an ethical matter.

One threat for the Urapmin seemed to be that chaos might break out were people to start talking freely about one another's intentions and thoughts. The issue is not what is psychologically possible but what the consequences of losing control over these possibilities are. Again, to worry about this is to cast a cognitive capacity in ethical terms, a question not of what one *can* do but of what one *ought* to do. Under the opacity doctrine, children have to learn not to express other people's thoughts and to monitor their own self-expression. This learning, it would seem, is precisely what gets undone when the Urapmin convert to Christianity and start to confess other people's thoughts. In these cases, the issue is not whether thoughts can be put in words but who can do it. The opacity claim is responding, in part, to the asymmetry between verbalizing my thoughts and verbalizing yours. It might be taken as an assertion of the right to be the first person of one's own thoughts and an acknowledgment of others' right to be the first person of theirs. Christian confession commonly aims to bring out a very special class of inner states pertaining to specific notions of sin and personal agency (you are not, for instance, supposed to confess to accidents). Confession shifts the first-person point of view to a third-person one. It stages you, the speaker, as split between the person who performed the misdeed and an observer who reports it. Confession thus teaches you how to speak with the authority of the first person in public—to put that authority into words and not fear the consequences.

The anthropologist Rupert Stasch (2008) points out that statements by Korowai about not presuming to impinge on one another's self-determination are normative models for the political terms of people's coexistence. Opacity, then, is a component in how they conceive of interiority as the locus of personal autonomy. We can compare this to the idea of the conscience in Western political thought. An important strain in the Western intellectual tradition holds that conscience, being immaterial, is the one site that cannot be subject to external power. You can jail my body, but you cannot jail my conscience. In certain respects this idea finds a parallel in the Melanesian cases; both are political reflections on the experience of inner thought.

One of Stasch's examples is the receiving of a gift that seems to have no motivation or explanation. Korowai treat this as something that is due to unknowable thoughts. But an unexplained gift is surely never left entirely uninterpreted by the recipient. There seems to be something disingenuous about claims that I have no idea why someone gave me a gift. After all, gift recipients are also gift givers, and gift givers, we know across the ethnographic record, are commonly skillful manipulators, perhaps most skillful when they themselves are least self-conscious about it. So the more general principle here is perhaps again that opacity claims are not necessarily epistemological or psychological claims but, rather, expressions of anxiety about the role of other people in one's own life and freedom of action. Notice that the Chewong and Inuit, discussed above, also worry about what others are thinking and the effect on what one can do. But inverting the Melanesian stance, Chewong and Inuit apparently presume that they do know what others are thinking virtually in principle.

ONE'S OWN THOUGHTS

What is distinctive about the "first-person" perspective, according to the philosopher Richard Moran (2001), is that it puts one in a position to speak on behalf of, or take responsibility for, one's thoughts. The Melanesian opacity claim, when it takes the form of unknowing knowing—that I have no idea why people do what they do (yet I do)—is a way to acknowledge this. It explicitly disclaims a responsibility for what only another person can rightfully avow. More than this, it seems that Melanesian opacity claims, as a subject of cultural elaboration, can affect *self*-knowledge by taking this unknowability of others as a model of one's relationship to one's own thoughts. This is hinted at in the idiom of the inner theater in which one's heart speaks and one's self acts in response to those promptings, which suggests the possibility that one's own intentions and thoughts might drift apart from one's actions. This thematizes the general human capacity for self-alienation and self-deception. According to the evolutionary biologist Robert Trivers (1991), this capacity for self-deception has a function in social worlds, since it makes one's efforts to deceive others all the more persuasive, as those efforts will seem quite sincere. In ethnographic terms, much depends on the particular ways such general human capacities are objects of reflection and manipulation, in what contexts, and to what effect.

There is evidence in some Melanesian societies of persistent anxieties about self-alienation. For instance, Stasch reports that Korowai worry that they may fall victim to love magic or be overwhelmed by grief. Both love and grief show the disabling effects of other people on oneself. The dangers posed by love or grief, in this context, seem to hinge on underlying norms of self-containment. For people such as the Korowai, the loss of self-knowledge might be due to ambivalence, confusion, un-acted-upon desires, or unbidden ideas. Viewed in this light, the opacity doctrine sharpens Korowai people's attention to the possibility of self-alienation by insisting on the distinctiveness of the first-person position, by publicly laying claim to it. In doing so, the opacity doctrine links the ability to control what is hidden and what is to be revealed to broader norms of social interaction and problems of political power. It takes the experience of having inner thoughts, something the psychologists might say is universally human, as an affordance for conceptualizing quite specific social problems and practical responses to them. For Korowai, some of those problems are due to their highly dispersed and isolating settlement patterns, the ideals of individual autonomy associated with them, and the specific kinds of social tensions to which they lend themselves. Following Stasch, the ethical significance of the distinctions between self and other matter to Korowai because of their bearing on that freedom of action, which some anthropologists have argued must be central to any ethnographic grasp of ethics (e.g., Faubion 2011; Laidlaw 2014; Mattingly 2012).

The key point is this: if we treat Korowai claims that they cannot intuit another person's thoughts to be merely one expression of a particular worldview, we might learn something about human diversity and creativity. But if along with that we also recognize that this denial flies in the face of something that Korowai (and everyone else) show evidence of actually doing, we may learn more. And what we would learn is not just the banal lesson that humans are much alike but something more interesting. Denying a universal human propensity for intention-seeking, the Korowai opacity doctrine draws attention to and conceptually elaborates certain familiar experiences of inner thought, interactive miscues, secrecy, and deception. It brings these *ubiquitous* experiences to bear on quite *particular* ethical and political problems. Viewed in this light, the opacity doctrine's denial of "Theory of Mind" is in fact a way of reflecting on intention-seeking as a way to think about and respond to the specific ethical challenges of Korowai social existence.

LOCAL THEMES, AFFORDANCES EVERYWHERE

Consider another version of an opacity doctrine. According to Eve Danziger, Mopan Maya disregard questions of intent when gossiping or punishing misdeeds. Inadvertent transgressions are punished just as if they were purposeful ones. Moreover, judgment about what people say "is based exclusively on the perceived truth value of expressions and not on the intentional or belief states of the speaker" (Danziger 2006: 260). In other words, Mopan do not discriminate between errors of fact and outright lies—a distinction that turns on speakers' intentions and states of belief (see Sweetser 1987). Yet Mopan are quite capable of meeting the usual psychological tests of Theory of Mind and easily make statements about what other people "want," "believe," and "know." In Danziger's analysis, the key issue has to do with the ethics of speech itself, a sacred morality. This morality centers on a specific view of the relations between words and the world, that "linguistic words and expressions are considered to be related to their signifiers in ways that . . . transcend the volition of those who use them" (Danziger 2006: 262). As with Melanesians, it is not the case that the Mopan have no capacity to read minds or invent fictions: rather, these capacities serve ethical thought, leading to emphatic denial of something they are in fact doing. The denial in this case takes the form of an ontological claim about the nature of words. To reiterate, if Theory of Mind and intention-seeking are common to all humans, how these get played down or emphasized can contribute to quite divergent ethical worlds. Elaborated in some communities, suppressed in others, these cognitive capacities appear as both sources of difficulties in their own right and affordances for ethical work. We can learn more about the specificity of these communities by noticing this than by treating their ideas about other minds as the creations of self-contained cultural processes at work in an utterly distinct plane.

Korowai and Mopan opacity doctrines reach us as reports from distinctive cultural worlds. Ethnography has typically treated each of these worlds as a historical product constructed within a particular social order. But for the last generation or so, anthropologists have been arguing that we should not understand cultures or societies as bounded, self-contained, or internally consistent entities. For many of them, the old models of culture and society have lost their explanatory purchase. Does that mean that the apparently distinctive histories, ontologies, and ethical sensibilities of Chewong, Inuit, Sumbanese, Korowai, Mopan, Poles, Kluane, Americans, Daoists, Confucians, and believers in karma are ultimately superficial, that we can quickly pass through them to get to the natural essence of ethical life? As should be apparent by now, this book argues that we are not presented with a stark choice between simple alternatives. The domain of social interactions described here is one mediating step between the panhuman sweep of psychology and the contingent particularities of social history.

Rather than seeing cultural or social orders as containing moral rules that are manifested in norms of interaction, the observations I've touched on here point to an ethics embedded within interaction and the commitments on which interactions depend. These commitments need not rise to any actor's awareness. They draw on basic psychological capacities and propensities such as intention-seeking and reciprocity of perspectives, as well as on the coordination techniques of interaction such as turn-taking, much of which remains beneath the threshold of ordinary attention, especially when things flow smoothly. In this chapter we have glanced at some of the ways in which particular communities have picked out certain distinct aspects of interaction, giving them names and drawing attention to the ethical possibilities and challenges they might possess. But what more general bearing does the momentary flicker of mutual regard (or disregard) have on a sociologically embedded, culturally articulated, and temporally durable sense of ethics and its historical trajectories? This is the topic of the chapters that follow.

CHAPTER 4

Ethical Types

MORAL BREAKDOWN?

In the previous two chapters, I tried to show how people, in their mutual dependence, may draw on affordances they discover in cognition, psychology, and the local patterns of interaction to reflect on and elaborate ethical problems and possibilities. For instance, the basic human propensity for intention-seeking can become an object of fascination or anxiety and accordingly be elaborated (as in some Western contexts) or suppressed (as in parts of Melanesia). Although people may also invoke social rules or norms, the ways they do so are products of interaction as much as they are guides for it. I will have more to say about that in this chapter and those that follow.

Interaction works most smoothly when people are more or less unself-conscious about the patterns, habits, and expectations it involves. This chapter focuses on the processes that link practices of everyday interaction with ethical reflection. Recall the distinction between morality system and the encompassing category of ethics laid out in the introduction. A morality system centers on obligations that are supposed to be grounded in consistent principles of great generality. I suggested that these are commonly subject to a high degree of consciousness and are readily verbalized as rules and doctrines. This does not mean that reflexive awareness is absent from ethics more generally, but it does not necessarily play so dominant a role in every ethical world. There are morality systems designed to inculcate the unself-consciousness of habit—and conversely, the otherwise smooth flow of everyday interaction sometimes prompts one to step back and reflect on one's ethical life. As noted above, some morality systems dwell on the role of self-consciousness in ethical life. Some ancient Chinese philosophers, for instance, debated the paradox posed by trying to teach *wu-wei*, effortless action, since they took the hallmark of ethical authenticity to be spontaneity (Slingerland 2007). We might then ask how to locate the historically enduring and institutionally embedded systems in relation to tacit intuitions and habits. What I will attempt to do in the rest of this book is show how we can grasp their relations. These relations work in two directions. Morality systems draw on features of interaction. The ethics of interaction in turn responds to the precepts made available by morality systems.

I have argued that the dynamics of social interaction form a major instigation for giving accounts and making ethical claims for oneself in the course of everyday life. This process is an important factor in the development of ethical self-awareness. To be clear, ethical life does not *require* self-consciousness. But because explicit vocabularies and arguments endure over time, circulate socially, and are salient targets for elaboration and criticism, they play a crucial role in the social histories of ethics. Now, before proceeding, consider an alternative view of ethical awareness, what the anthropologist Jarrett Zigon (2007) calls "moral breakdown." Zigon remarks that "most people consider others and themselves moral most of the time, and for this reason it is rarely considered or consciously thought about." He claims that "the need to consciously consider or reason about what one must do only arises in moments that shake one out of the everydayness of being moral" (2007: 133).[1]

How does this account differ from the picture I am developing here? First, it assumes that the distinction between consciousness and unconsciousness maps onto that between distinct moments of moral decision making and the ongoing flow of habit and routine. The evidence I adduce here indicates that this is overly simple. Ethical self-awareness is not just a matter of decision making, nor is it confined to singular moments. Second, moral awareness is not just a matter of the inner thought of the individual confronted with a dilemma—much like the person found at the center of the trolley problem—unless we include the crucial element of interaction

[1]We will look at the somewhat related idea of "problematization" in chapter 5. For the distinction between the concepts, see Laidlaw 2014: 110ff.

with others. Drawing on intensive fieldwork in playgrounds with young American girls, linguistic anthropologist Marjorie Goodwin provides an alternative picture of ethical reflection. She finds that when they negotiate the rules of their games "across an array of different social groups, girls are extraordinarily adept and articulate in producing moves that explicate a sense of justice. Girls display intense engrossment in formulating logical proofs and demonstrations for their positions and exhibit determination in the pursuit of their positions" (2006: 246). Here we see clashes to be a familiar occupation of everyday life, induced by interactions with others. Finally, the idea of moral breakdown assumes that "the primarily goal of ethics . . . [is] to once again dwell in the unreflective comfort of the familiar" (Zigon 2007: 138). In effect, this reproduces one traditional view of culture, in which people move more or less smoothly along the grooves that have been laid out for them. But if ethical life is not defined by conscious deliberation, neither is it limited to habit and intuition (cf. Lempert 2013). On the one hand, ethical pedagogies from Aristotle to early Confucians and Daoists to Islamic piety movements (see chapter 6) have aimed at habituation to virtue, and ethnographic evidence certainly shows that a vast amount of ethical life *is* spontaneous, habitual, or routine. On the other hand, people's capacities for reflection, criticism, and even alienation are also ubiquitous parts of human life and should not be treated as rare or peculiar. Ethics is not all of one order. Sometimes people are in the midst of the action; sometimes they seem to stand apart from it. There are many circumstances and motives that might induce the effect of standing apart or prompt reflexivity. The approach to ethical life sketched out here aims to keep the variety and motility of these different stances in view. We will return to the sources of purposeful efforts at consciousness-raising in chapter 5.

In this chapter, I will continue to work outward from the evanescent details of conversational interaction toward the enduring, and sometimes globe-spanning, systems of morality that are most familiar in public discourses surrounding ethical life. The core argument that links these chapters is that neither the fleeting moments of ethical staging and judgment in the flow of everyday life nor the pious and principled claims of moral philosophies can be fully understood without grasping their relations to one another. Sometimes they are set in opposition to one another; sometimes they work together. In this chapter, we look at how people take ethical stances and make claims for themselves or others, in more or less explicit ways. It asks in what situations (whether interacting with others or even in solitary contemplation) one is called upon to give an account of oneself.

This chapter continues to hone in on the close analysis of people talking, but with the aim of shifting attention toward the more expansively ethnographic and historical materials that take up the final part of the book. This chapter begins by looking at how people take ethical stances in conversation. *Stance* here means "a way of categorizing and judging experience particular to a group or individual" (Kockelman 2004: 129). The evaluations are public, intersubjective, and embodied rather than private, subjective, and psychological. Moreover, as the linguist John Du Bois (2007) observes, normally the individual alone does not take a stance, since stance involves simultaneously people's relationships to each other and to some third entity to which they are attending. Evaluative stances saturate interaction through and through. They are reflexive, judging not just the situation but the self, and that reflexivity has implications for the other participants as well. In what follows we will look at some examples of stance and then consider the ways in which those stances can take the form of "voices" that embody locally recognizable types of ethical persons or actions.

SELF-AWARENESS AND OTHER PEOPLE

To start, here are some fragments of conversation. The first comes from an interview with a young man at a British rock concert in the 1980s, explaining to a researcher why he dressed in a Goth style:

> I mean ever since I started wearing clothes which I chose then they may be conventional clothes but I probably wear them in a strange way you know in a strange mixture so they always so I always you know look strange to people anyway even if I was wearing conventional clothes and it went on really cos feeling different to everybody

> else and what I'm wearing is a way of expressing the isolation you know if if you felt isolated from everybody else then once you get to the age where you can choose for yourself then I would choose to wear clothes where I I could show what I felt differently to everybody else around me so that's how I started doing it. (Widdicombe 1993: 104–5, quoted in Antaki 1994: 131)

In looking at this quite banal passage, the analyst points out the work the speaker is doing to establish that his choice of dress is personal but not identical to the supposedly personal choices a mere consumer might make in the course of shopping, because unlike mere shoppers, his choice is authentic. In the first line, he shifts from the past tense to the present and makes a concession that seems to anticipate a possible objection by the listener ("they may be conventional"). According to Widdicombe, the shift in tense shows that the speaker is dissatisfied with the way he started. He quickly realizes that he might give the impression that he is just following fashion. It seems that he is hearing how he might sound to the interviewer as he speaks and tries to forestall what he realizes could be a likely reading of his words. However isolated this man feels himself to be from the people around him, he is highly attuned to how he looks from their point of view. Moreover, by insisting on his authenticity, he seems to assume that his listener shares with him certain values such as independence and authenticity. Or at least he uses the perspective afforded by that listener as a way to reassure himself. If we take his insistence on his authenticity to be an ethical claim about his sense of virtue, it is inseparable from the way he gives an account of himself, both verbally and sartorially. In this respect, ethics depends both on the presence of other people and on the semiotic forms—in this case both linguistic and sartorial—that mediate relations between them and him.

An even briefer fragment illuminates just how ubiquitous and ordinary the ethical staging of the self is. It comes from a study of calls to an emergency service phone line in Britain:

CT [OPERATOR]:	Mid-City Police an' Fire
C [CALLER]:	Hi um (.) I'm uh (.) I work at thus University Hospital and I was riding my bike home tanight from (.) work

CT: mm

C: 'bout (.) ten minutes ago, 'hh As I was riding past Mercy hospital (.) which is a few blocks from there 'hh ( ) um I *think* uh couple vans full uh kids pulled up (.) an started um (.) they went down thus trail and are *beating* up people down there I'm not sure (.) but it sounded like (something) 'hh (Zimmerman 1992: 438–39, quoted in modified form in Antaki 1994: 134)

Here the analyst draws a stylistic contrast between this way of reporting a crime and a simpler approach, in which the caller simply says, "'hh Um: yeah (.) Somebody jus' vandalized my car."

How do we account for the difference? According to the researcher, the caller in the first example is trying to establish him- or herself as an ordinary citizen with a job, who is a neutral witness and not part of the action being reported and whose report can be trusted because it is punctilious about what he or she can know for sure. Perhaps because the report is not necessary, since the caller is not directly affected, unlike the owner of the vandalized car, some further positioning seems called for. Transcripts such as these hint at how saturated ordinary exchanges are with the ongoing business of establishing one's ethical worth in the eyes, or ears, of others, the everyday task of giving an account of oneself.

When Butler says that one is called upon to give an account of oneself in an encounter with an accuser (see chapter 2), she evokes a scene whose structural parallels run through the ongoing doings of everyday life. In the examples just above, the speaker addresses someone as a potential accuser from a (minor) position of unequal power. Little encounters like these can be effective in constructing ethical realities precisely because, saturating experience, they are so easy not to notice. More noticeable but still commonplace is gossip. Marjorie Goodwin (1990, 2006) has shown how complex this activity can be, by analyzing the genre of "he-said-she-said" carried out by working-class African American girls in Philadelphia. In Goodwin's description,

> a girl accuses another of a particular infraction: having talked about her behind her back. The offended party confronts an alleged offending party because she wants to "get something straight" . . .

> [for instance, saying,] "And ***Tan***ya said that ***you*** said that (0.6) I was showin' off just because I had that ***bl:ouse*** on." . . . These declarative utterances establish the accuser's ground (warrant) for the accusation, how she learned about the offense. Responses to the accusations are typically denials ("***Uh*** uh. I ain't say anything") or accusations about the intermediary party's work in setting up the confrontation ("Well she lie. I ain't ***say*** that"). Indeed, within a single utterance a girl can invoke a coherent domain of action, a small culture, one that includes identities, actions, and biographies for the participants within it, in addition to a relevant past that warrants the current accusation, and makes relevant specific types of next actions. (2006: 7)

Here are three distinct ethical offenses identifiable with three accusers and defendants: the topic of the original gossip (showing off), the original act of speaking (malicious gossip), and the act of passing that speech along (he-said-she-said). These girls are implicitly giving accounts of themselves and of others: as show-offs, as gossips, and perhaps as traitors. They have ready-to-hand an array of ethical categories they can make more or less explicit, about kinds of person and types of action. These are not necessarily subject to prolonged reflection, but they are available in the face of accusations: each girl could, if called upon, give an account of herself and of her playmates.

In contrast to Butler, who seems to invoke a Kafkaesque scene in which the speaker is summoned to stand before an overpowering figure of the law, to face judgment and possibly dire punishment, these examples give us quite trivial stuff indeed. Yet there is a structural resemblance. One does not simply account for oneself because one is endowed with the spirit of inquiry. Something must instigate the giving of an account. It could be the accusation of an authority or, say, the religious rituals of confession and absolution (Carr 2013). But self-accounting does not require intensely loaded circumstances, which (if one is lucky) might be quite rare. It might just arise from efforts to be seen as normal (Sacks 1984) or, in some subcultures, *not* normal (Warner 1999). Both the Goth and the crime witness are responding to a demand, actual or imagined, by someone else whom they take to be evaluating their words and whose evaluation matters to them. Moreover, that evaluation seems to take place with reference to some values implicitly shared between speaker and addressee. In the first case, the Goth is striving not to be taken as conventional and superficial but to be recognized as authentic; in the second, the caller tries to be seen as normal and therefore reliable. In either instance, the consequences may be incidental—perhaps in both cases the main risk is simply not being taken seriously. But the power of such maneuvers lies in their very ordinariness and ubiquity. Saturating the ongoing flow of everyday life, they prompt certain kinds of self-awareness. They respond to the presence of other people; the intersubjectivity they involve is driven, in part, by people's ongoing efforts at ethically evaluating others and themselves.

STANDING BEFORE THE LAW

The minor but ubiquitous expectation that people will display regard for one another in these ways can be taken to be the basis for ascriptions of character. Thus conversations are not just about ethical judgments (as in gossip); the technical requirements of interaction also may themselves afford inferences about intentions or character. It is a commonplace of sociolinguistics that even barely perceptible differences in timing, prosody, or word choice can lead people to intuitive character judgments, that, for instance, one's interlocutor is arrogant or timid, honest or crooked, macho or effeminate, deferential or insensitive, endowed with gravitas or hopelessly trivial (Gumperz 1982; Sacks 1972; Tannen 2005). What Sacks wrote decades ago hardly seems dated in 2015:

> Among the Americans, the police are occupational specialists on inferring the probability of criminality from the appearances persons present in public places.... Patrolmen are intensively oriented to the possibly improper appearances persons may present ... [starting] with the fact that persons within the society are trained to naively present and naively employ presented appearances as the grounds of treatment of the persons they encounter in public places. (1972: 282–83)

Social interactions cannot work as smoothly and quickly as they do without people relying on a vast number of default assumptions on the basis of which they draw inferences about who their interlocutors are and what exactly is going on. Those inferences are typically extremely fast and below the level of awareness. They are not inherently ethical, but as these examples suggest, the coordinating devices that facilitate the mutual orientation of participants to one another and the reciprocity of perspectives afford the making of ethical judgments. And, as we will see in subsequent chapters, they may provoke more deliberate ethical reflections.

The major task of conversation analysis has been to demonstrate the organizational principles by which interactions are constructed. The approach often treats the question as a problem in traffic management—how you know when it's your turn to speak, who gets to take the next turn of talk, how you signal that it's time to stop talking, and so forth. But these technical requirements also afford evaluations. Goffman captured this long ago, writing,

> When a stranger comes into our presence, then, first appearances are likely to enable us to anticipate his [social] category and [personal] attributes, . . . such as 'honest.' . . . We lean on these anticipations that we have, transforming them into normative expectations, into righteously presented demands. (1963: 2)

Note the shift from the quasi-objective descriptive mode of the social category to the ethical language of righteousness.

This sense of normative expectations is especially salient for someone who is stigmatized on grounds of ethnicity, profession, sexual orientation, or disability, since the presence of "normals" means "having to be self conscious and calculating about the impression he is making, to a degree . . . which he assumes others are not" (Goffman 1963: 14). The stigmatized person may be induced to see him- or herself as a *type* of person, as seen from the perspective of others, the third-person point of view—rather than as the subject of first-person inner experience or the second-person addressee of an immediate interaction between particular individuals. One consequence can be "stereotype threat" (Steele and Aronson 1995). This refers to the lower performance of people who have been prompted to think of negative stereotypes about their own social group. But Goffman suggests that since the possibility of stepping outside of the action and taking a third-person point of view is a potential built into social interaction more generally, "normal and the stigmatized are not persons but perspectives" (1963: 138). Thus the stereotype threat, apparently, can strike even members of a dominant social group, for example, when whites are experimentally induced to worry about appearing racist (Frantz et al. 2004).

One of the schoolgirls studied by Goodwin became so ostracized that her parents considered moving to another street. There are of course more powerful judges. Anthropologist/social worker E. Summerson Carr studied very poor women in a drug treatment program in an American inner city. The program, which controls their access to social welfare services and even custody of their own children, is geared to evaluating the addicts' progress, which relies on indirect measures of their propensity to relapse. To convey the stakes of talk in this fraught and highly asymmetric relationship, Carr presents the reader with a thought experiment:

> Imagine you are in exile in a foreign land and have been diagnosed with what the natives consider to be an incurable, if treatable, disease. This disease is characterized by the inability to use language to express what you think and how you feel. You are now being treated by local specialists who work to rehabilitate your relationship to language. . . . [T]he specialists judge the extent to which your utterances match your inner states or—in native terms—how "honest, open, and willing" you are as a speaker and as a person. *Honesty*, *openness*, and *willingness* are of the highest cultural value; they are the indigenous markers of individual integrity, morality, and health. . . . Since the specialists' evaluative powers are linked with the capacity to distribute basic goods and resources, your ritual performance—that is, the way you speak in the course of your treatment—has far-reaching material and symbolic consequences. (2011: 1)

Because addiction is considered to be a chronic condition, all one can hope for is that some addicts will become sincere adherents to the treatment. But sincerity, being itself an inner condition, can only be known by its external signs. Since those signs are perceptible, the client can, in effect, stand outside herself and see them as the social worker does. Once the client learns those signs, the possibility of mimicking them cannot be ruled out. This diagnostic problem is compounded by local ethical theories. The social workers, and many of the clients, believe that sincerity and honesty are ethical virtues in their own right. Therefore, to fake it or "flip the script" is not only to produce false diagnoses; it is also to fail as an ethical subject altogether.

This possibility is implicit in the nature of social interaction. Self-presentation must be legible to others. But that very legibility opens up the possibility of seeing oneself from their point of view as well—it is grounded in the reciprocity of perspectives developed early in childhood, as we have seen. Although ethics does not necessarily depend on self-awareness in any given moment, it seems to be thoroughly entangled with things that can produce self-awareness, such as the use of explicit ethical categories in talk. As we will see in the next section, the play among these perspectives can be both complex and productive of ethical reflexivity.

THE INNER CLASH OF ETHICAL VOICES

The linguistic anthropologist Jane Hill (1995) writes about an elderly man she calls Don Gabriel. Don Gabriel lives in an Indian farming village in central Mexico, near an industrial suburb of Puebla. He is a speaker of the Native American language called Mexicano or Nahuatl, which is being flooded out by Spanish. Hill recorded a long monologue in which he recounts the murder of his adult son nine years previously. The son had been caught in the ongoing conflict between the values of the villagers and those of the dominant society around them. Relations between village and surrounding society are complex, but villagers tend to portray them in terms of a stark ideological opposition. On the one hand, there are the values of mutual respect, community solidarity, and economic reciprocity among Indians engaged in subsistence agriculture; on the other, the drunkenness, vulgarity, competitiveness, and individualism of urban capitalists. The former ethical world is identified with the Mexicano language, and the latter, with Spanish; both languages permeate Don Gabriel's speech. Don Gabriel's son, having taken a job as treasurer of a local bus service against his father's advice, was seen by villagers as betraying local values for those of capitalist accumulation. He became the target of *envidia*, a destructive form of envy, which led to his murder. In telling his story, Don Gabriel is caught between his identification with his son and his identification with the village values his son was seen to betray.

In a close analysis of the transcript of Don Gabriel's narrative of the event, Hill argues that it stages a contest of the two value systems. Sometimes Don Gabriel uses explicit ethical vocabulary, such as *desgraciadamente* (unfortunately), and "loaded" words, such as *ambición* (ambition) and *ventaja* (advantage), to evaluate what is being narrated (Hill 1995: 118). But most of the ethical contest does not take explicit propositional form. It is revealed in linguistic and stylistic variations in Don Gabriel's speech. These include switching between Mexicano and Spanish; dysfluencies such as false starts, stutters, and verbal slips; shifts in verb tenses; and changes in intonation contour. In teasing out the patterns among these stylistic variations, Hill identifies a set of distinct "voices." *Voice*, in this sense, refers to the identification of a speaking style with a kind of person (Bakhtin 1984; Voloshinov 1986). It is an elaboration of a phenomenon noted in my review of early child development in chapter 1, that internalized voices can be a source of moral authority (Haidt 2001; Tomasello 1999; Wertsch 1991). To this fundamental cognitive capacity, Hill's analysis of adult speech adds two observations. First, internalized voices can be expressed outwardly by drawing on the range of linguistic features available in a given speech community. Second, those voices may represent either momentary stances or more enduring conventional personae—what Agha calls "characterological figures" (2007: 165)—that may be quite distinct from the speaker's sense of identity. For example, irony and sarcasm function precisely by underlining the distinction between the speaker's words and his or her own commitments. Or a speaker may take on a voice identified with a locally recognizable figure, such as the nagging mother, the canny salesman, or the innocent child. In principle, any perceptible feature of speech can serve to indicate a certain voice. But as we will see, what gives those variables meaning is their contribution to the speaker's evaluative stances. For voice not only depicts an ethical position; it also depicts the speaker's stance toward it, for instance, indifferent, doubtful, approving, or critical. Even apparently routine activity involves a range of ways of taking stances, from fully inside to completely alienated, with degrees of self-awareness and conflict along the way. This variability is quite ordinary; no particular "moral breakdown" is required.

According to Hill, Don Gabriel's voices embody by turns a neutral narrator, an engaged narrator, an emotionally overcome protagonist (the murder victim's father), and a moral commentator on the events the narrator is portraying. The latter addresses the listener directly, using historical present verb tenses and certain intonation contours that emphasize immediacy. Hill argues that Don Gabriel's speech shows him working to position himself within the range of ethical stances in this community, a task he is not fully able to master. For example, Don Gabriel uses only Spanish vocabulary such as "treasurer" (*tesorero*), "surplus" (*sobra*), and "savings" (*ahorro*) when referring to the dealings for profit, as if he could draw moral boundaries around the community identified with the Mexicano language. Trying to keep his own distance, Don Gabriel even puts criticism of commerce into the mouths of others, as when townspeople say, in mixed languages, "que timotehuiah īpan in cargo por base de interés" ("that we fight one another over the job due to personal interest") (Hill 1995: 134). Yet, even then, he tries to assign this vocabulary to voices at a remove from the moral center of his narrative; when a voice close to that center speaks of finance, it is in euphemisms: "nīnqueh cristiano, in āquin ōquipiayah ambición de nōn, . . . de nōn trabajo" ("these people, who had ambition about that, . . . that work") (Hill 1995: 135). According to Hill, the Spanish lexicon of profit "remains for him alien, and his struggle with it [is] a principal source of dysfluency in the otherwise eloquent flow of his narrative" (1995: 108). Don Gabriel's deployment of voices seems to be part of an effort to gain recognition from his interlocutor as being a good person. He may also be making an ethical claim on that interlocutor, asking for collusion in the construction of his own self-image (bearing in mind that Don Gabriel may himself be a crucial audience for his own ethical performance). In verbal interaction, voices call for a response of some sort, an affirmation, for example, or cooperation in consolidating or gaining recognition for the position being voiced.

DYSFLUENCY AND ETHICAL CONFLICT

Hill's essay is of particular interest because it is influenced by an approach to social interaction that, as noted in the previous chapter, Alasdair MacIntyre thought peculiarly *amoral*. But for Don Gabriel, moral values are precisely what are in question. Although he is clearly attentive to how he is presenting himself to an interlocutor, the appearances he is trying to manage involve serious ethical commitments and the difficulties that their contradictions pose for him. In Hill's analysis, stylistic variations deploy both a cast of social types who themselves manifest distinct commitments and visions of the good and the speaker's identification with or estrangement from them. The key moments hinge on dysfluencies, which Hill takes to be evidence of underlying trouble that interrupts what would otherwise be a smooth performance. Hill offers us an intriguing alternative to some other accounts of the failure of speech in the face of trauma, such as E. Valentine Daniel's (1996) description of people who cannot speak at all when they try to recount violent episodes in Sri Lanka's civil war or Veena Das's (2007) analysis of victims' loss of voice after massacres and rapes in India. Don Gabriel's failures are quite different:

> Passages having to do with the death are among the most artful and fluent in the narrative.... Don Gabriel can speak the death of a child. What resists his voice and "falls out" of his fluent narrative art is the language of business and profit. (Hill 1995: 137–38)

This is not the willful and amoral self that MacIntyre ascribes to Goffman's world. To the extent that Don Gabriel has trouble controlling the voices he utters, his talk resembles that of a spirit medium, who is possessed by a spirit but does not thereby surrender complete control to it. Possession is a struggle in which the medium always has a say. Don Gabriel's encounter with the ethical universe is neither created tabula rasa nor scripted in advance but draws on the raw materials at hand. His words do not just express a self that stands wholly independent of them, for it seems that his self-knowledge depends upon those very voices.

According to Hill, variations in verbal form offer clues to two things, the presence of multiple voices in the narrative and a struggle for dominance among those voices and the ethical positions they index. In choosing a stylistic option, Don Gabriel is not simply trying to present a pleasing appearance to an interlocutor or to make the interaction run smoothly. Rather, it seems here that the presentation of the self to a listener is a kind of ethical work on the self. It may even be part of a discovery process by which Don Gabriel works out his ethical commitments and which contradictions he cannot master.

But because the discursive resources on which the speaker draws are inherently dialogic in two senses—they both juxtapose social possibilities and are addressed to other voices—work on the self cannot easily be a single-minded process. According to Hill, Don Gabriel tries to establish a coherent ethical position among conflicting ways of speaking, weighted with contradictory ideologies, by distributing them. For example, he cannot simply eliminate the presence of the voices of the profit seekers—among other things, he would be unable to account for his own son's actions—but he can locate them as far as possible from the voices that express the ethical stances he wants to affirm. Yet these voices continue to interrupt his efforts at control. For just as he participates in a social world that includes profit-seeking, and is not a wholly monolithic ethical actor as the idealized portrayal of the egalitarian farmer would have it, so too the voices that index that world remain part of his own discursive repertoire and cannot be entirely silenced. Thus Don Gabriel's self-discovery depends on other people whom he addresses or invokes and the linguistic resources available to him in doing so.

Don Gabriel's voices may exemplify something endemic to the ordinary ethics of everyday life. Hill focuses on the way in which speakers move among different identifiable voices as they take up the various ethical possibilities at play within a community. To capture this, she quotes Mikhail Bakhtin, whose writings about nineteenth-century Russia reveal even small-scale social life to be a matter of struggle among distinct ideological and normative positions:

> An illiterate peasant, miles away from any urban center, . . . nevertheless lived in several language systems: he prayed to God in one

> language (Church Slavonic), sang songs in another, spoke to his family in a third, and when he began to dictate petitions to the local authorities through a scribe, he tried speaking yet a fourth language. (Bakhtin 1981: 295–96)

Bakhtin's observation goes beyond the truism that communities are internally conflicted, for the play of voices does not merely take place among individuals. If Hill's analysis is right, it shows that this play occurs within the speech of an individual. Don Gabriel is not just following a cultural script. Hill's analysis suggests not only that ethical conflict may be endemic even within a small community but also that it may contribute to the ethical self-formation of its members.

DISCIPLINING THE CLASH OF VOICES

The clashing voices of Don Gabriel all inhabit a single monologue. We can't know for sure how consciously artful or not his performance is—it is probably an unstable mix of both. This talk may represent one moment in his emergent ethical self-awareness. But if so, this consciousness seems not to be the result of moral breakdown or moral dumbfounding but, rather, something both more uneven and more ubiquitous in everyday life. Some characterological figures, such as that of the virtuous Mexicano-speaking farmer, are quite stable and easily spotted by speaker and listener alike. Others, such as the voice of judgmental comments on the side, may appear and disappear again, lacking a stable identity attached to a recognizable verbal style.

Were the distinctive voices at play in his talk to become fully identifiable, attaching verbal styles to social stereotypes, they could enter into the public staging of ethical figures. Crystallized as generally known ethical types within a given community, voices could be purposely appropriated for didactic ends. This is exactly what goes on in the moral education of diasporic Tibetan monks analyzed by linguistic anthropologist Michael Lempert (2012). One of the speech genres Lempert discusses is the harangue of the disciplinarian, addressed to the gathered monks. It may be triggered by the lapses of certain individuals, but it is carefully addressed to no one in particular. The "penal semiotics" here, as Lempert puts it, make use of the full range of formal properties of speech, such as grammatical devices, word choice, reported speech, metrical contrasts, repetition, vocal quality, and volume, in order to produce a clash of voices.

The didacticism is not just a matter of stating things; it is also produced through the forms with which the disciplinarian expresses himself. By highlighting ethical distinctions, the clash of voices renders them more salient and sharpens the listeners' awareness of them:

> In reprimand, the disciplinarian seeks to reform the moral dispositions of monks through projecting, juxtaposing, and evaluating morally weighted social voices, like the voice of the "derelict monk" . . . whose failings are spelled out, sometimes in graphic detail, enough to make the guilty parties squirm; whose subjectivity is exposed to the glare of public scrutiny, but whose voice remains anonymous. No finger-pointing, no proper names. In explaining his art to me, the disciplinarian praised this brand of indirection. The objective is to induce in the wrongdoer's mind the thought "I am the one" about whom the disciplinarian speaks. . . . Rather than resort to the blunt instrument of biographic individuation, the disciplinarian populates his discourse with morally weighted voices and invites his audience to consider them and adjudicate between them. Like Hamlet's staging of the play *The Murder of Gonzago*, the disciplinarian delivers a stinging morality play that depends on voicing for its effectiveness. (Lempert 2012: 110–11)

The effectiveness lies in part in the manipulation of address in order to construct an ethical figure—the derelict monk—in such a way that anyone at all might identify with it:

> This indirection . . . speaks to all because it is addressed to no one. . . . That is, there is a certain infectious paranoia that the disciplinarian's low-resolution sketches of social types can induce; with enough scrutiny, you begin to wonder, Is he talking about *me*? You recognize yourself in these nebulous portraits. (Lempert 2012: 114)

The disciplinarian not only stages the voices of virtuous and derelict monks. He also stages a virtuous detachment from his own apparent anger:

> For those who would mistake his affective displays as signs of vulgar worldly "anger" (*zhe sdang*), the disciplinarian diagrams his subjectivity

> as dissimulated, as split into (external) wrathful affect and (internal) benevolent motivation. . . . Part of the way he communicates dissimulation and indexes this category of affect is through an anonymous, indirect addressivity. In reprimands, Sera Mey's disciplinarian uses language to refer to infractions and moral failings, but without naming offenders. (Lempert 2012: 110; citations omitted)

In contrast to the previous examples, here the speaker is highly self-conscious about his goals and the semiotic means of achieving them. His way of speaking is quite purposefully oriented toward the production of character types that are already known and nameable. It presumes the existence of a morality system of explicit obligations and prohibitions. In the case of these Buddhist monks, of course, this morality system has doctrinal roots that reach deep in history and are sustained by elaborate institutions, rituals, and the more or less centralized authority of the Dalai Lama. These are topics to which we return in the next chapter. But first, we need to look more closely at the nature of voices in actual practice. Here I want to consider two other things that Lempert's study shows us. One is the enactment of character; the other, the role of virtual observers.

Tibetan monks regularly engage in a highly stylized and dramatic style of debate. As one monk defends a point of doctrine, others flamboyantly attack him, in what looks like emotional and potentially physical violence; they "stamp their feet, hurl taunts, and deliver loud claps," often right in the face of their opponent. Confronted with these attacks, defendants are supposed to remain utterly impassive, "publicly marking themselves as *kinds* of people endowed with culturally valorized characterological attributes" (Lempert 2012: 50). Moreover, the resilient character they display goes beyond personal virtue; it also displays the idealized character of the doctrinal texts they are defending: "If the defendant is convincing, he mimics how tradition ought to be: stable, coherent, whole" (Lempert 2012: 50). Through the perceptible demeanor of the individual, personal character is identified with the morality system, whose qualities in turn are, in some degree, made available to the concrete experience of participants and other witnesses.

Not all witnesses are physically present. According to Lempert, a set of virtual witnesses—which are themselves ethical stereotypes—haunts contemporary Tibetan Buddhism. Tibetans are aware of "a host of distant critics, which include such named figures as 'scientists,' 'foreigners,' 'Communist China,' and 'Mao Zedong'" (Lempert 2012: 4). As a result, they are adept at seeing themselves as others see them, from the third-person point of view, and shape their actions accordingly. Now in certain respects, as we saw in the previous chapters, this capacity to see oneself through others' eyes is a fundamental feature of ordinary intersubjectivity. What is distinctive about the situation here is the sense Tibetans have of entering a historically novel scene in which their own needs for political recognition and support are imposing on them a new set of ethically marked expectations. In this respect, their relation to those third persons is one condition for potential ethical revolution, the topic of a later chapter. But first, one more look at ethical voicing and its relationship to third persons, now in a kind of scene many readers may recognize.

TYPIFYING CHARACTER EXPLICITLY

Lempert tells us that a Tibetan monk wins a debate both by what he says and by the kind of person he displays himself to be. Such displays are meant not just for the immediate opponent but for third persons as well, both members of the immediate audience and imagined critics in distant places. All of this might seem quite familiar to followers of American electoral politics. In this light, let's take another look at the 1996 presidential debate between Bill Clinton and Bob Dole discussed in chapter 2. Agha points out that the participation structure of the debate extended well beyond the two immediate protagonists and the moderator, Jim Lehrer. There was an audience in the room, as well as television viewers and radio listeners, political commentators, reporters for the print media, and (after the event) their readers. Although the debaters addressed one another directly, obviously in an election campaign the most important audience is found in these more distant and diffused publics. These publics also provide a useful way of registering the locally available ways of evaluating and describing the ethical qualities of social interactions and individual character. Agha plays off the transcript of the debate itself against how it was reported in an attempt to tease out how people recognize veiled aggression.

Near the start of the debate, Clinton, who at that point was a sitting president running for reelection but already touched by scandal (notably the so-called Whitewater and Troopergate cases), says, "I want to begin by saying again how much I respect Senator Dole and his record of public service, and how hard I will try to make this campaign, and this debate, one of ideas, not insults" (Agha 1997: 476). Clinton refers to himself in the first person three times; he even describes his own act of speaking as he does so ("I want to begin by saying"). Like the girls engaged in the "he-said-she-said" we looked at above, here Clinton explicitly characterizes his own speech (respect, trying hard, ideas not insults). Put another way, this is a highly self-conscious act of ethical positioning that draws on a widely recognized vocabulary of character and stance that, precisely because it is a vocabulary, is easy to point to. Notice that Clinton praises Dole in the same sentence as his claim that he himself will avoid insults. As Agha observes, what could easily be reported by the press was the "compliment." What was harder to point to was the way that juxtaposing the two statements gave the latter coercive force. Naming Dole in the same breath as his reference to the eschewing of insults, in effect, serves as a demand that Dole do likewise.

At this time, one of Clinton's political vulnerabilities was precisely the "character issue." The challenge that Dole faced, then, was how to keep the matter of character in the fore while not appearing to violate the call to keep to the high ground. According to Agha, he largely accomplished this through acts of veiled aggression, which, while avoiding explicit insults, used context, parallel constructions, and other discourse features to convey the same effect. For example, Dole also makes an explicit reference to respect while simultaneously calling into question the character of the very person he claims to be respecting: "And it seems to me that uh there's a problem there Mr. President. And I will address you as Mr. President. . . . You didn't do that with Mr.—with President Bush in 1992" (Agha 1997: 478). The clash is between the explicit metalanguage of respect, character, and ethical action, on the one hand, and the implicit actions within which that metalanguage is embedded. Moreover, each act of self-characterization ("I am respectful") can, in this context, be taken to imply an opposed characterization of the addressee ("you are *not* respectful") (Agha 1997: 480). But the latter, merely an unspoken shadow projected by what was actually said, remains a form of aggression hard to pin down.

As in the examples of the Goth, the caller to the police hotline, and Don Gabriel, a great deal of work is going into ethical self-characterization, which (to a greater or lesser degree) functions in coordination with the characterization of others (the Goth is authentic, others are mere fashion followers; the caller is innocent, the thugs are not; Don Gabriel is virtuous, some people are envious or greedy). In one way or another, all of these are little scenes in which one gives an account of oneself, something undertaken for an addressee or witness. This is certainly the case in the Clinton–Dole debate as well. Where that debate differs from them, and resembles the example of Tibetan debate and discipline, is in the additional dimension contributed by explicit ethical language. Clinton, Dole, and the commentators *name* and *evaluate* what is going on. In this case, of course, the candidates are following their calculations of political strategy. Being explicit, moral language is relatively easy to manipulate for ulterior purposes. This is one reason why ethical life can never be adequately captured by vocabulary items alone. But explicit language and the circumstances that call for its use are often crucial elements in the production of those morality systems that have public recognition and durability, as we will see in chapter 6. First, however, we need to pull together some of the threads that have been laid out here.

ETHICAL FIGURES AND TYPES

The voices staged by Don Gabriel and the Tibetan disciplinarian, and the explicit self-characterizations performed in the Clinton–Dole debate, sustain ethical evaluations. They embody, or portray, the speaker's stance toward something. As noted above, *stance* refers to evaluations in terms of some shared value, as they are staged within a social interaction. The individual alone cannot take a stance, since, in what the linguist John Du Bois (2007) calls "the stance triangle," stance involves people's relationships to one another and to something else that they are evaluating, such as an action, a remark, or a person. Du Bois notes that this parallels the emergence of intersubjectivity described by Tomasello (1999), in which young children learn to align themselves with others with respect to some third object of attention.

In the stance triangle, two or more individuals align themselves with each other with respect to their respective evaluations of an object, toward which they are also positioning themselves as subjects. Yet the intersubjectivity of stance does not stop there, since it takes place in time. As a result, "the very act of taking a stance becomes fair game to serve as a target for the next speaker's stance" (Du Bois 2007: 141). As we saw in the case of Mr. Taylor, the unfortunate missionary we met in chapter 2 who was accused of drunkenness, and in Philadelphia girls' gossip, the stance attributed to us on the official record, consolidated over the long term, is ultimately out of any individual's hands (Irvine 2009). This is partly a function of the sheer temporality of interaction. But it is also shaped by larger-scale social phenomena such as institutional power, prevailing stereotypes, and even the constraints built into the media of interaction. For example, your words are more easily taken out of context, and your chances of making corrections are usually more limited, in writing than in face-to-face conversation.

Don Gabriel's voices point to different ethical values and stage evaluative stances toward them. This works because stylistic variations can be identified with social categories or kinds of person. But any given formal features of speech cannot inherently point to anything in particular (Hanks 1992). To identify a stretch of discourse as a "voice," it must be linked to some characterological figures that are recognizable to the participants. These figures must be known beforehand or have the potential for being known in retrospect—that is, when brought into play, they might induce in the listener an "aha!" moment of recognition. The figure is thus repeatable and capable of circulating across distinct contexts. Once a certain way of speaking becomes recognizable in different contexts, it helps produce the effect of a person having a certain knowable type of character, of a certain ethical inflection and socially identifiable identity, and then it takes on the full-fledged nature of a social figure or stereotype.

If Hill is right, Don Gabriel's voices index value systems that in some respect preexist his invocation of them. On the other hand, by giving voice to ethical values, by putting them into circulation, he is contributing to their consolidation and endurance within the community and potentially transforming them as well. Each performance may, potentially, continue to stabilize the ties between verbal style and ethical type or figure, these voices becoming easily available for appropriation and circulation by other speakers. In this way, to those who recognize them, the speech style and social type may come to seem to be bound together quite naturally. Don Gabriel's voices form a bridge from the immediate here and now of face-to-face interaction to the concepts, values, and conflicts that may have very long histories in his community. This bridge leads in both directions. Voices are only recognizable if they evoke types, ethical figures that are actually or potentially known in other contexts beyond *this* particular moment of interaction. They do not vanish at the end of any given conversation. Conversely, those figures' reality depends on their recurrent invocation in concrete interactions and on their being recognizable by other persons. Like ancient gods, once people cease to make them offerings, they cease to exist.

The voices and figures that Hill finds in Don Gabriel's speech manifest a dynamic that the sociologist Alfred Schutz (1967) calls typification. This refers to an aspect of the routine processes that make habitual life possible. In Schutz's analysis, although you experience your intimates in highly individual terms, you also constantly interact with people about whom you need to know little more than their social type. For example, you may need to know how to interact with a cashier or police officer in their professional roles but not as someone's father or sister. Put this way, *typification* merely refers to a process that underwrites people's everyday experience of one another. But types are rarely just neutral categories like "retiree" or "clerk." As the objects of people's stances, they are prone to be evaluated. As Manning's depiction of interactions in Starbucks suggests, even the barista–customer relation can be ethically fraught. If *stance* sometimes refers to judgments that flit across moments of interaction, when typified and objectified in the form of characterological figures, stance can also involve values whose force is amplified by their recurrence from one context to another. The analysis of "voice" indicates one bridge between general features of intersubjectivity, discussed in chapter 1, and figures produced within historically specific morality systems, the topic of the final chapters.

DEFINING THE SITUATION

Earlier chapters of this book argued that ethical life cannot be simply a matter of decision making. It should be equally apparent that it is not confined to stance-taking. This chapter has focused on giving accounts as a matter of interactions with others. The giving of accounts is characteristically provoked when people need to allocate responsibility for an action. This may occur not only in public negotiations, formal accusations, or didactic discourses but also in the ordinary and ubiquitous flow of conversation. Recall that Butler describes giving an account of oneself before an authority able to punish the speaker. In a characteristically minor key, the philosopher J. L. Austin (1979) also discusses the giving of accounts when he asks, "When do we excuse conduct?" His response focuses on conversational interaction:

> In general, the situation is one where someone is *accused* of having done something, or (if that will keep it any cleaner) where someone is *said* to have done something which is bad, wrong, inept, unwelcome, or in some other of the numerous ways untoward. (1979: 175–76)

Notice the role of speech here, that talk is doing two things that depend on each other. On the one hand, this talk singles out a person or persons as responsible, but, on the other hand, it must also describe a situation or an action as one that calls for judgments about responsibility, even things it makes ethical sense to have emotional responses to, against the implicit possibility that the situation does not call for ethical evaluation.

One of Austin's types of excuses involves how one characterizes an action. He points out a distinction in English-language usage between *justification* and *excuse*. A justification focuses on the action. Confronted with a possible misdeed or transgression, the work of justification depends on redescribing or reclassifying the action so that it is not a wrong. A killing done in battle, being justified, is not murder (Austin 1979: 176–77). An excuse, on the other hand, focuses on the agent: a killing arising from a road accident on slippery ice, being unintentional, is not murder (at least in the society in which Austin writes). So, we can rephrase Austin's remark that "it is to evade responsibility, or full responsibility, that we most often make excuses" (1979: 181) to say that descriptions of actions and persons in terms of ethical categories are provoked by accusations, actual or potential. These descriptions may draw on locally available ideas about causality. In societies such as Sumba or the Mopan Maya, where people are likely to think that there are no accidents, even an icy road might not serve as an excuse. Elsewhere, the agent's intentions might be irrelevant: Oedipus punishes himself for acts he knows he did not intend.

Consider here another fragment of the transcription mentioned in chapter 2, of a recording made while two women, roommates in Mexico City, broke out in a bitter quarrel after they had each been simmering for days:

13 L; OH YEAH?
14 P; YOU HAVE SEEN ME ON LOTS OF OCCASIONS
15 L; THIS IS THE FIRST TIME
16 P; BECAUSE I GO TO ANOTHER HOUSE
17 AND YOU ARE INVOLVED IN YOUR OWN AFFAIRS, WHICH SEEMS TO ME . . .
18 L; SO THIS IS THE FIRST TIME THAT I HAVE TALKED TO YOU? (HAVILAND 1997: 550; original Spanish transcription omitted)

What is going on here? P has returned from a trip involving difficult personal matters and is angry that L has not asked her about it. According to Haviland,

> On one possible reading of P's extended turn in lines 16–17, she has heard L's *es la primera vez* "this is the first time . . ." in line 15 as the

> preface to an excuse. . . . She thus understands L to be saying something like, "we haven't talked, despite the fact that we have seen each other many times, because this is the first time we have . . . (had the opportunity to talk? been in circumstances that would allow us to touch on personal themes?) . . . ." P rushes in to cut off and thus to discount such an anticipated excuse, obscuring an attempted "turn" by L even as she *interprets* it. (1997: 560)

Let us assume that the quarrel focuses on the responsibility of one roommate toward another, to show concern, offer an ear, to be there for her. L seems to be offering an excuse that preserves her virtuous standing as a responsible roommate. In order to forestall this implicit self-characterization, P rejects her definition of the situation.

If the action itself is not a transgression, then ethical responsibility can be a moot point. But there's more to the question than dodging accusations. The typification of actions lies at the heart of certain definitions of ethics itself. Within his tradition, the philosopher Charles Taylor (1985) argues that one can only be fully responsible for action as one knows (or should know) it. That is, a precondition for the attribution of responsibility is the establishment of what Goffman (1959) called a "definition of the situation," Anscombe (1957) called "action under a description," or Silverstein (1993) called "metapragmatics." But there are no categories of action simply out there in the world. As Velleman puts it,

> We talk about "taking" an action, as if we were picking an apple from a tree, but actions don't antecedently exist in nature, waiting to be picked. What we call taking an action is actually *making* an action, by *en*acting some act-description or action concept. Which actions we can make depends on which descriptions or concepts are available for us to enact. (2013: 27)[2]

Even the choice of medium may determine how a situation is to be construed. In her study of American undergraduates' use of electronic

[2]Some psychologists claim that people do pick out the boundaries of actions on the basis of their innate abilities to recognize other individuals' intentions (Baird and Baldwin 2001), but even so, this still seems to leave the *characterization* of the action indeterminate.

media, Gershon notes that "sometimes, even when one person [in a relationship] believes a breakup is final, the other person does not accept its finality until it is repeated in another medium" (2010: 398). To break up with someone by text or e-mail might be taken as just an implicit summons for further discussion; by contrast, to do so in the more public domain of Facebook can be definitive.

What is key here is to recognize that the giving of reasons and the ethical description of actions are hardly confined to the grave debates of high theorists—they run through the most banal moments of everyday life. That very ubiquity is part of what gives them their power. Here is a minor yet, I think, wholly characteristic example of an especially self-conscious moment of didactic objectification. It's a transcript recorded just after a kids' pickup basketball game in Los Angeles, when the father invited the two teams to cheer each other:

> FATHER: DOES ANYBODY know what that's called?
> REGAN: Yeah.
> FATHER: What?
> REGAN: Good game.
> FATHER: Good sportsmanship.
> FATHER: It's not only about . . .
> REGAN: Let's play another game. (*does cartwheel*)
> FATHER: It's not always about winning. It's about being a good sport.
>
> Father asks the kids to label the behavior that they have just exhibited. Regan suggests "good game," echoing part of the utterances used in the cheer. Father corrects Regan that the action represents "*good sportsmanship*." (Kremer-Sadlik and Kim 2007: 44; special transcription markings removed)

Here the father gives the flow of action a description (being a good sport) and anchors it with explicit reference to a larger ethical theory (good sportsmanship). This small moment of objectification, with its almost ritualized dialogue structure, is not radically different in kind from more formal activities such as debates about justice (Dave 2012), sermons (Hirschkind 2006), or catechisms (Keane 2007)—or even from the telegraphic metalanguage of a New York crack dealer: "Real crazy. Yeah! Ray's a fuckin' pig; Ray's a wild motherfucker. He's got juice. You understand Felipe? Juice! . . . On the street that means respect" (Bourgois 2003: 24). And all share certain features with gossip, which in Zinacantan, as perhaps most places, "is clearly *aimed* at the interlocutor. People try, through gossip, to convince their interlocutors, to arouse their sympathies, or to recruit their support" (Haviland 1977: 83–84).

In all these cases, the explicit verbalization of ethical descriptions is prompted by an encounter between, respectively, children to be taught, political opponents to be won over, congregations to be admonished, believers to be disciplined, ethnographers to be wised up, and neighbors to collude with. These examples of ethical talk seem to require that the people be able to see themselves from outside, taking a third-person perspective on their own thoughts and actions as instances of a recognizable category.[3] But this kind of reflexivity also depends on the objectification that results from interactions with others. This objectification (in the first example above, the words *good sportsmanship*) in turn endows it with at least the potential for historical endurance beyond the moment. Thus people may continue over time to take "good sportsmanship" to be a standard by which they can judge themselves and others, perhaps as a virtue peculiar to particular social orders or a model for international goodwill like, say, the Olympic Games. We will return to the expansion and circulation of ethical categories in the final chapters of this book.

INTERACTION AS AFFORDANCE

The technical details of conversation largely work beneath participants' awareness. But when linguistic forms or patterns in interaction become recognizable, cognizable, and repeatable, they can take on the status of public objects, basic elements of a morality system that can be invoked from the third-person perspective. In Samoa, for instance, "hierarchy is maintained through the display of *fa'aaloalo*

[3]To say this does not mean that "description" is limited to discrete vocabulary items; the grammar of modality, for instance, is another resource. Austin (1979: 187) remarks that excuses are often adverbs, the assertion that, yes, one did do a certain action—but only in a certain way.

(respect), a term composed of the causative prefix *fa'a* (to make) and the respect-vocabulary term *alo* (to face) thereby entailing a bodily and social orientation toward others" (Ochs and Izquerdo 2009: 397). These orientations seem to draw on some basic physical and verbal dimensions of interaction grounded in the early childhood development of joint attention and primary intersubjectivity (Baron-Cohen, Baldwin, and Crowson 1997; Trevarthen and Aitken 2001). Samoans have brought them into focus by rendering them explicit ethical objects, knowable values that people can talk about explicitly. These provide the raw materials for representational redescription. As noted in chapter 1, this is a cognitive skill that developmental psychologists say the child must acquire in order to be able to take up the third-person perspective necessary for full-fledged social life (Karmiloff-Smith 1992; Tomasello 1999). Moreover, these objects of knowledge reinforce that "sustaining social contribution to character" to which Merritt (2000) refers when defending virtue ethics against situationist psychology. Samoan interaction inculcates knowable values:

> Family members rarely issued praise. It was also unusual to compare individual skills in family surroundings. . . . Rather, one performing an activity was viewed as benefiting from a *tâpua'i* (supporter). Distributed responsibility was recognized through a common verbal exchange: A tâpua'i notices a person's activity by saying "Mâlô!" (Well done!), then the person reciprocates "Mâlô le ta apua'!" (Well done the support!). Through such exchanges children were socialized to give and receive support and that responsibility resides beyond one person. (Ochs and Izquerdo 2009: 398)

Interaction is not a one-way process, a simple inculcation of values, since it gives rise in turn to further interpretation and reflection. In Zinacantan, the person being told gossip must actively push the narrator for clarifications and details:

> The nature of the interaction makes it possible for both interlocutor and storyteller to emphasize the relevant aspects of the affair; the gossip session serves as a practical moral lesson by allowing participants to reflect on particular behavior and observe its outcome before

> making explicit evaluations. . . . [As a result] people build ethical theories on evaluations of such situations. Zinacantecos, through gossip, continually test ordinary rules and evaluative words against actual behavior. (Haviland 1977: 54–55)

The reflexive activities of gossip do not simply reproduce values but exert new pressures on them. These and the other basic aspects of human cognition and interaction reviewed above have the potential for becoming foci of attention. Subject to reflection, they can give rise to practices and ideas that, as we will see below, are both long lasting and subject to transformation.

A critical aspect of the oratory of the Tibetan disciplinarian is his use of forms of speech that address no one in particular. Lempert observes that this invites each listener (in this all-male audience) to take himself as a potential addressee. In this way, the very forms of talk help build up the sense that the orator is giving expression to a general moral order, something that transcends the immediate setting and any particular offenses anyone might have committed (Lempert 2012: 114). The disciplinarian is taking advantage of contingent, individual transgressions to produce a generalized "we." As we saw in chapter 1, the generalized we and the shared intentionality it expresses build on some very general capacities for reciprocity of perspectives and third-party norm enforcement apparent in early child development (Bratman 1992; Kalish 2012; Moll and Tomasello 2007; Tomasello et al. 2005). Formalized as a device in the hands of the Tibetan disciplinarian, these cognitive capacities are given specificity and mobilized for purposes of inculcating the kinds of social institutions that endow any particular morality system with its claim to generality and permanence. Much of the rest of this book concerns the relationship between such general moral orders and the more intimate ethical phenomena we have been looking at so far. The cloth of ethical life is woven of threads drawn from *both*: explicit concepts and rules, as well as tacit habits and intuitions.

In sum, verbal interaction is not merely an arena within which character is established or challenged, recognition offered or denied. It is also the preeminent site where people may demand explicit reasons and accountings of one another or provide them. It is in response to the demands posed by talk that rationalizations and justifications arise. The natural home of argument, reasoning, and justification is not in the individual autonomous mind but in palpable social interactions, whether face-to-face or in more mediated forms—for even doctrinal texts imply an addressee. In the next chapters, I will also argue that these practices can result in the objectifications that endow ethics with its historical character, something that endures beyond the momentary situation but can also change beyond recognition.

PART THREE

Histories

CHAPTER 5

Awareness and Change

Shifting Stances

In the previous chapter we encountered the idea of stance, which refers to one of the ways in which the flow of ordinary interaction is saturated with evaluations. Most of the time, these evaluations are not just momentary or idiosyncratic. To repeat words quoted in chapter 4, "Which actions we can make depends on which descriptions or concepts are available for us to enact" (Velleman 2013: 27). This chapter asks what it means for a description or concept to be "available."

I hope by now to have made clear two things that "being available" does not mean. First, the descriptions and concepts with which people understand their actions are not simply built into human nature as facts of cognition or psychology. But neither are they simply impressed upon them by some cultural mold or pure social construction. They are the product of interactions between shared human capacities and the diverse ways people have reflected on and responded to them. Since any moral vocabulary is both shared and subject to historical changes, so too are the resources on which people can readily draw in making those everyday evaluations that social interactions prompt. This is especially the case when conceptual vocabulary and ethical stances are made explicit, as people draw on them to give accounts of themselves, offering praises, justifications, excuses, or condemnations to one another. As they do so, they may wittingly or unwittingly be exerting pressure on, or discovering new aspects of, existing norms and habits.

Consider here some changes in English semantics since the eighteenth and early nineteenth centuries. Don Herzog draws our attention to the word *condescension* as it was used in that earlier period:

> In 1832, John Payne Collier was working in the Duke of Devonshire's library when the duke brought him his lunch. A duke performing the duties of a servant: the occasion was notable and Collier recorded it in his diary. "He always does his utmost to lessen the distance between us, and to put me at my ease, on a level with himself," he mused. "I do not call it condescension (he will not permit the word), but kindness." (1998: 206)

A similar use of the word appears in a letter written in 1780 by one Gamaliel Lloyd to a member of Parliament: "We are all greatly obliged to you for the trouble you have taken in giving us your sentiments, and for the condescending deference you pay to the opinions of your Constituents" (quoted in Herzog 1998: 206).

As Herzog remarks, spoken in England or America today, a sentence like the latter could only seem sarcastic, since among presumptive equals condescension is arrogant and patronizing. But in 1780, condescension was a virtue, "the act of a great man who graciously lowers himself to deal with inferiors on a footing of equality" (Herzog 1998: 206). So the very notion of condescension is incoherent without the background of deeply embedded assumptions about social hierarchy. Condescension in the eighteenth and early nineteenth centuries was the voluntary act of temporary self-lowering of a superior to a social inferior. Like the Samoan vocabulary of respect (*fa'aaloalo*) noted in the previous chapter, the feelings people seem to have had in eighteenth-century England, along with the virtues and ethical stances they embodied, are simply not available to those who lack the background social realities they presuppose. A middle-class American in the early twenty-first century cannot reproduce this sense of open condescension, at least not in a way that would be recognizable and acceptable to his or her next-door neighbor, because neither of them inhabits a world in which overt social hierarchy is an accepted ordering principle.

The point here is twofold. On the one hand, the descriptions under which people can understand their actions, and the evaluative stances those descriptions entail, depend on aspects of the context that extend well beyond the immediate situation in which they occur. To feel or to recognize condescension cannot be merely a direct expression of the emotions or an objective reckoning of a situation. The very existence of condescension depends on a host of habits, expectations, and concepts that the participants bring into the situation. These habits, expectations, and concepts are shared with people outside the immediate interaction. This is in part because they are embedded within the institutions, the political economy, and the folk psychology that sustain the hierarchy they reflect. When the duke brought lunch to Collier, another eighteenth-century Englishman would have been able to make sense of the situation and understand Collier's response, possibly recalling other similar incidents.

On the other hand, these habits and concepts are not stable. Just a few generations later they have vanished, lost even to the immediate descendants of those who once took them for granted. Moreover, there were already hints of instability at the time in which they were thriving. For one thing, condescension can only gratify its recipient if everyone agrees to the initial terms of the hierarchy. Even in an era when the superiority of dukes could be taken for granted, those conditions might fail for other individuals in other contexts. Here's Samuel Johnson writing about a visit to an old friend he calls Prospero, who has grown rich:

> The best apartments were ostentatiously set open, that I might have a distant view of the magnificence which I was not permitted to approach; and my old friend receiving me with all the insolence of condescension at the top of the stairs, conducted me to a back room, where he told me he always breakfasted when he had not great company. (quoted in Herzog 1998: 208–9)

In this case, we might say that Johnson could not accept the hierarchical claims of someone he once knew as an equal. Prospero did not merit the relative status he claimed for himself—but someone else, such as a duke, for instance, might. Johnson is not attacking the moral category of noble condescension as such, only its misapplication in this instance. Prospero was, in fact, guilty of "insolence." This is another word whose eighteenth-century meaning has been lost, since it likewise presupposes an accepted status hierarchy, in light of which it is unwarranted presumption. Like condescension, insolence can only make emotional or conceptual sense in a world in which there is some consensus about those status differences and a general willingness to acknowledge their inequality openly.

Twenty-first-century Americans are certainly unequal in many ways, but they lack a shared moral vocabulary for settled hierarchy: I cannot consider you to be condescending to me, in that earlier sense, unless I first accept your legitimate superiority to me. The key here is the ethical value implicit in that legitimacy: that the duke is my superior, it is fit and right. (This sense of fitness might be easier to grasp for today's reader, who might find the superiority of dukes too hard to swallow, if we replace the nobles with something even more exalted. One of the earliest attestations of the word in the *Oxford English Dictionary*, an essay on the Lord's Prayer written in 1676, addresses God Himself: "Give us a sense of thy Great Condescention to thy weak and sinful Creatures.") As the moral economy has changed, so have the possible ways of describing a person's character and actions and (or so Herzog argues) the very emotions those actions can prompt and possess.

But the matter is even more complicated. It would seem that by the time Collier was writing, there were already competing evaluations on offer. Egalitarian values were struggling with hierarchical deference, to which alternatives were imaginable. Thus, in the first anecdote, although Collier seems to have wanted to call the duke's action "condescension," the duke rejected this description. Why? Herzog remarks that even at the time, it was not necessarily viewed as an unmitigated good. An act of condescension does not eliminate inequality; it only stages an image of equality, one that is temporary and in which all the agency remains on one side. Apparently, merely seeming equal was not necessarily acceptable to those being condescended to. And acceptance is crucial, since one cannot be condescending unilaterally. Herzog underlines the interactive character of these terms:

> None of them can get off the ground without two parties. You need someone else to condescend to, to be insolent to, to be impudent to.... Each refers centrally to the social interaction between the pair, not to any feelings either is experiencing. And each sometimes refers only to a relatively detached account of the action: one can condescend or

> be insolent or impudent without intending to, without noticing that one's actions might be interpreted that way. (1998: 217)

The point is crucial. Although the language of condescension and insolence is fraught with emotional resonance, and may be tapping into innate intuitions about fairness and hierarchy that emerge early in all children's development, this vocabulary refers above all to the quality of relations *between* persons, not to the psychology of their *internal* states. In part, this is what makes these terms not just emotional, or even (as Herzog is arguing) political, but ethical.

They are ethical because they manifest the ways people evaluate one another and depend on one another for recognition of their own value. Collier and Dr. Johnson cared about the condescension they received in part because each saw himself through the eyes, respectively, of the duke and Prospero. But they were not telepathic: all of this was mediated semiotically, through interpretable actions. The tray, the lunch, the library, the words spoken, the bend of the noble torso as it placed the tray before the scribe—all situated this interaction between individuals within a larger, knowable, social world. This moment of recognition was endowed with its value not just by an individual with a tray, or a man at a desk, but by everything that embedded them in the social world of other dukes and other scribes. As that world changed, the very same actions—carrying a tray to serve lunch to a visitor, accepting a constituent's petition, receiving one's old friends in the back room—would come to fall under very different descriptions or perhaps hardly be noticeable at all.

But can changes in such small matters of etiquette and insult really amount to the stuff of serious history, ethical or otherwise? How do we locate the minutiae of everyday social interactions in the grand scale of social histories as they are usually portrayed? The period Herzog discusses runs from the era of mad King George III, through that of the French Revolution, and on to the parliamentary reforms. Somewhere in the background of these anecdotes, however distant, we might hear the arguments of people such as Edmund Burke and Tom Paine about equality, tradition, and the Rights of Man. We may also hear traces of the popular sermons of preachers during the religious revival called the Great Awakening. Certainly the background assumptions that give sense to condescension and insolence are affected by new social conditions, changing economic relations, and innovative vocabularies. But one of the main ways they enter ethical life is through the everydayness of ordinary interactions. In this way basic psychological features such as the child's propensity to seek fairness encounter historically specific concepts and institutions, such as the nineteenth-century English idea of justice. This is not a unidirectional process. Ethical concepts such as justice or equality can only be fully realized if they become part of the everyday life of ordinary interactions. This imposes constraints: after all, the history of ethical ideas is full of great aspirations and dramatic proposals that, in the end, came to nothing. Not every moral system can take root, certainly not in every circumstance. Some, like that of the Shakers, are so radically austere that they are unsustainable.

The period Herzog describes saw the demise not of social hierarchy as such but, rather, of its easy acceptance. Its defenders found themselves anxious and defensive. They were increasingly caught in contradictions that made wholehearted emotional commitment to inequality hard to legitimate. A few generations later, the meaning of *condescension* has become so altered that it can be hard to persuade our contemporaries that it once held positive connotations. Should individuals wish to act with condescension—as, no doubt, people do (bankers being solicitous of their housecleaners, perhaps)—it cannot be carried off with the same open confidence that their superiority is accepted. In this respect, social history has altered the emotional resources we know as psychology. But are these changes random, or do they have a direction?

ETHICAL PROGRESS?

It is obvious that, at some level, ethical life does change. Human sacrifice is vanishingly rare. Although slavery may not have disappeared from the planet, it is no longer widely accepted as normal and natural. People in places as far apart as Jakarta, Nairobi, Los Angeles, and Buenos Aires can speak of universal human rights, support gender equality, and reject forms of cruelty to animals in ways that might have been unimaginable just a generation ago. Some of them are engaged in forms of self-cultivation, introspection, and individualism that would have seemed strange or even repellant to their grandparents. Yet many naturalistic and normative approaches tend to treat ethics as something fundamental and universal. But if ethics is simply built into the human condition, whether we understand that claim in, say, biological, cognitive, or sociological terms, then how could ethics change? How can we reconcile the claim that ethics rests on secure foundations with the observation that it has a history? Here are three common answers: (1) the content of ethics is affected by the growth of scientific knowledge, (2) the conceptual character of ethics progresses from rules to principles, and (3) the domain of ethics expands in scope as people's social worlds grow.

The first answer parallels some common narratives about the rise of science. In this story, information is cumulative, and knowledge eliminates errors. As humans discover more about the natural and sociological world, they learn not to succumb to common mistakes. Some of these mistakes concern the nature of reality. If once people sought morality in divine mandates, and blamed misfortune on occult forces, now they know better. They seek good and evil in brains, philosophical principles, or social arrangements (Neiman 2002). Sometimes this process can seem to shrink the domain within which ethics is relevant. As the anthropologist E. E. Evans-Pritchard (1937) pointed out long ago, people who believe in witchcraft can explain more than scientists can. They can tell you why an event that from a scientific perspective is the indifferent conjunction of distinct causal processes (a termite-ridden granary collapses on me just at the moment I happen to be napping under it) has a reason—it is the purposeful work of a malevolent witch. Once you no longer believe in magical agency, certain events lose their ethical point. The granary collapsed due to the action of termites (something the believers in witchcraft also know); but now my presence when it falls is mere chance.

Another version of this first answer concerns progress in our understanding of the psychology of ethical judgments. Some writers draw an analogy to optical illusions: although I cannot stop seeing heat waves on the road as water, knowing that this is a mirage, I can compensate appropriately (Appiah 2008: 110; Bloom 2004: 191).

Similarly, some psychologists and other social scientists argue that our growing knowledge of human cognition and emotions will allow us to compensate for the misperceptions and biases described in chapter 1. Knowing our resistance to pushing one man in front of a runaway trolley in order to save five lives, we may compensate for the effect through the utilitarian calculations of a rational judgment. In this view, scientific knowledge can have a therapeutic effect on ethical life.

Again, the result can be a shrinking of the ethical domain. What appeared to be acts of judgment might turn out to be the effects of something quite different, such as a neurological process. This is what happens when the kinds of neurological and psychological research discussed in chapter 1 are taken up socially. The anthropologist James Laidlaw points to one example, a legal case involving a man named Charlson who had murdered his young son. Charlson's lawyers persuaded the court that the killer had a brain tumor that rendered him not responsible for his actions. According to Laidlaw, "The defence depended on the conceptual detachment of the tumour from the 'self' that was the defendant, on the idea that it was a distinct, detachable, hostile, and intrusive object that was not part of him" (2014: 193). Notice here the appeal to the third-person perspective and the subordination of the killer's own subjectivity to that of mechanical explanation. This is the very kind of sharp distinction, between the world of causality and that of evaluations, of ethical self-knowledge, on which some radical reductionists (e.g., Harris 2010, 2012) claim to have proved that we should do without the idea of morality. But to be effective, that case depends on public criteria of acceptability, which are embedded in law and local theories of agency and causality. (Moreover, the result may not be exculpatory: as noted in chapter 2, on discovering the facts of the matter, Oedipus takes responsibility for crimes that he did not, from the first-person perspective, intend to commit.) Charlson, the man with the brain tumor, could not simply chose, all on his own, to decide that he was not the person who killed his son, at least with any ethical or legal effect. The tumor had to become part of his way of giving an account of himself, which had to be recognizable to those to whom he owed that account (in this case, a judge and jury). This second-person address seeks recognition that in turn depends, among other things, on a taking up of a third-person perspective on the event. Charlson, or his lawyer, had to offer a plausible account of the effects of the tumor that was not apparent in his own experience. Charlson had to be able to take himself as an object of reflection in order to detach himself from his action. Self-objectifying third-person reflexivity need not take the form of biomedical science, of course. It may be theological, as when a penitent confesses to God. Or it may be sociological, as when the hoods in the musical *West Side Story*, mentioned in the introduction, disclaim responsibility because their mothers are junkies.

The expanding realm of facts can seem to squeeze out that space within which ethical judgment had once been called for. But the results may also lead in the opposite direction too, as people become more alert to the full range of causes and effects in which they are involved. They may feel compelled to acknowledge new responsibilities, for example, when employers discover unintended biases in their hiring practices or consumers discover their connections to distant child labor or far-off climate change (Laidlaw 2010). Or they may discover the ethical implications of previously unself-conscious habit: as one sociologist of conversational interaction remarks, "My aim is to make visible (and thereby to enable us all to challenge) some of the mundane quotidian actions that result in the routine achievement of a taken-for-granted world that socially excludes or marginalizes non-heterosexuals" (Kitzinger 2005: 478). In either case, this shrinking or expansion exemplifies an assumption underlying much social science: the revelation of facts should dispel misjudgments of value. At the same time, the possibility that the outcomes might be diametrically opposed illustrates Laidlaw's observation that the ethical conclusions one draws are not simply given by the facts of the case, since "whether any causal account in fact illuminates and explains always depends on whom the explanation is for" (2014: 184).

This first version of ethical history, a story of growing knowledge, is also one about increased knowledge and control. After all, knowing more about optical illusions doesn't make them go away; it just allows one to become more aware of one's intuitive responses in order to thwart their effects. Now, it's not obvious that this story is substantially true. It would be hard to find a good metric by which to measure ethical progress as a function of growing scientific knowledge. The story's acceptability to contemporary readers may be partly due to its familiarity—it just sounds right. This might be because the story is similar to one that once dominated Western thought, the triumph of secular enlightenment. It begins with the retreat of religious superstition before the growing light cast by Protestantism and culminates in the greater light of science. In what I have called the moral narrative of modernity (Keane 2007), humans once lived in thrall to gods and spirits, onto whom they had projected the powers that in fact belong to themselves. In order to reclaim their own powers, humans had to overcome this displacement and recognize the true state of affairs.

The second story of progress recounts a shift in the style of ethical thought from rules to principles. Like the previous one, it centers on the effects of increased consciousness on human capacities. In this version of the story, however, what is growing is not people's knowledge of a factual realm but, rather, their grasp of underlying moral principles. It goes something like this: The history of morality begins with people following tribal customs, ancient traditions, or sets of rules handed down from elders, rulers, priests, or ancestors. They follow the rules more or less by rote, without much reflection. Their ethical lives are directed either by deference to the authority of those customs, elders, or gods or by fear of the consequences should one break the rules. Progress consists of becoming aware of one's own moral agency and therefore becoming able to make ethical choices. Those choices are based on having developed a grasp of the principles and justifications that underlie the old rules. One might, for example, replace the biblical Golden Rule with the Kantian Categorical Imperative.

These reflections enable people to liberate themselves from passive obedience and therefore to see themselves as responsible for their choices. Only in this way (according to certain traditions of moral thought) can they become fully ethical agents. They know what they are doing, and they are doing it for the right reasons, guided not by habit, acquiescence to authority, or fear of punishment but by a better appreciation of the innate worth of ethical values. By grasping underlying, and fairly abstract, ethical principles, they can make ethical decisions when encountering novel situations about which the customary rules might be silent. This is one reason why the ability to reflect on ethics, and not just act ethically, has been so central to many ethical theories (Laidlaw 2014; Parfit 2011; Schneewind 1998). Like the previous narrative, this too has roots in religious history, echoing the story that Protestants told about the liberation of the individual conscience, guided by thoughtful scripture reading, from blind adherence to Catholic rituals and priests. It is a certain way of telling a story about the growth of human freedom, linking ethics to choice, choice to agency, and agency to awareness. By stressing principles over concrete practices and customs, it also tends to separate morality from any ties to specific places and immediate social relations. As Godwin, whom we met in the introduction, advised, I should save the bishop rather than the chambermaid, even if she is my own mother.[1] The outcome of the narrative is an ethics based on abstract concepts that could apply to anyone in any place.

The third narrative of ethical progress can be called the expanding ethical circle (Singer 1981). Here what changes is not necessarily the content of ethics but its scope of application, the range of who counts as ethically relevant. According to this narrative, people's ethical concerns start out, historically, being limited to their most immediate social surroundings. In some cases, this means that notions of right and wrong simply do not apply to outsiders to one's group. Or they might apply to outsiders, but only when they are one's guests (Shryock 2012). In other cases, this narrative appears as a kind of relativism, the acknowledgment, for instance, that everyone is bound by obligations and prohibitions, but the precise rules are specific to each group. In Sumba, for instance, the meat of a certain animal is prohibited to one clan, and the water from a certain spring is off-limits to another. So too, God's covenant with Moses bound him only to certain people, not others.

[1]This delocalization is exemplified by the theory of justice developed by the philosopher John Rawls (1971), which proposes a thought experiment in which you must design a social system without knowing what position you might have in it, much the way one might ask one child to cut the cake and then give the other child first pick of the resulting pieces. This "veil of ignorance" is meant to guarantee an ethics of disinterestedness.

According to the expanding ethical circle narrative, this range tends to grow over the course of historical time, encompassing ever wider social groupings. It may, for example, begin with one's immediate kin, with whom one shares certain taboos and obligations, and reach to a tribe, a religious or ethnic group. With the rise of nationalism, ethical consideration may be extended to one's fellow nationals at the prompting of state institutions. Another important impetus comes from below. For example, through social movements, groups of people who had been excluded from ethical consideration may force more powerful groups to acknowledge their rights (Anderson 2013), a topic to which we will return.

The culmination in the narrative of the expanding ethical circle is universalism, whereby a true morality must apply to all humans or even beyond that, to include animals or the Earth itself. In the contemporary world we can see versions of this ethical universalism in proselytizing religions, the international human rights movement, and humanitarianism, to which we turn in the conclusion. This narrative can work in conjunction with the second, since the move from local to universal application seems to call for a parallel shift from specific rules to abstract principles. And like the second narrative, the expanding ethical circle has antecedents in religious polemics, such as those that celebrate the move from the special pact between Jews and God to the universal claims of Christianity.

As I have noted, each of these narratives has a religious genealogy, at least in part. Each has been used to justify the claims of one group over others. This should make us cautious—the very ease with which we can accept them may be due as much to their familiarity as to the power of the evidence in their favor (would, for example, the modern invention of scientific racism count against them?). So the task we face is to sort through the historical and ethnographic evidence. If ethical history has a direction, but is not guided by divinity, how can we account for it? If it does not, what drives the changes that clearly do occur?

Although much of this history is a tale of unnoticed causes and unintended consequences, it is also one of purposeful efforts at ethical transformation. That is the topic of these final chapters. In this one, we look at changes in people's awareness. It is a theme that runs through the narratives of ethical progress I have just sketched. For example, one story turns on people changing their minds about what entities exist in reality (witches? souls? angels? Logos? germs? neurons? genes? socioeconomic classes? rationality? national character?) and therefore shifting their ideas about where we should place responsibility. An alteration in people's picture of reality alters the descriptions of kinds of actions and kinds of agents that are available to them. People who no longer blame misfortune on ancestral spirits might instead worry about corporate responsibility (Welker 2014). As noted above, Hacking argues that descriptions of people can have "looping effects." Writing of the diagnosis "multiple personality" he says,

> In 1955 this was not a way to be a person, people did not experience themselves in this way, they did not interact with their friends, their families, their employers, their counselors, in this way; but in 1985 this was a way to be a person, to experience oneself, to live in society. (2007: 299)

As growing numbers of people found this to be a way to be a person, and came to the attention of doctors, their variety in turn began to alter the psychiatric definition. And as descriptions of actors change, so too do possible actions. By supplanting agents such as witches from consideration, germ theory may eliminate moral responses to disease in favor of biomedical ones. Secularized ideologies of language may no longer recognize the curse and the blood oath as serious ethical acts but, rather, matters of etiquette or jurisprudence (Keane 2009). Changing descriptions might in turn have political consequences, as the feminist Gloria Steinem points out when she remarks, "Now, we have terms like *sexual harassment* and *battered women*. A few years ago, they were just called *life*" (1995: 161). When the scope of ethical concern expands to include a wider sociological range—more people—it commonly involves a redescription of the world. In transforming *savages* or *gentiles* or *heretics* into fellow human beings one may also lose whole pantheons of ancestors, gods, ghosts, spirits, angels, saints, and other obsolete ethical subjects. So how does the world come to be redescribed, consciousness rearranged, and what are the consequences?

THE SOCIAL PRODUCTION OF ETHICAL PROBLEMS

As the earlier chapters of this book showed, important aspects of ethical life operate below or beyond people's conscious awareness. Some, such as cognitive biases, may be fundamentally unavailable for inspection. Others, such as the flow of interaction, may become visible, but work smoothly only if the participants do not pay too much attention to them most of the time. And still others, such as the background assumptions that pervade a given social world, are simply so taken for granted that the need to attend to them rarely arises.

Although words such as *condescension* and *insolence* may slip easily into the flow of everyday life, as they did in eighteenth- and nineteenth-century England, they do suggest that people are noticing something about the ethical quality of their interactions with one another and finding it worth remarking on. According to the *Oxford English Dictionary*, *condescension* is a neo-Latin coinage that seems to have come into use in the seventeenth century. So we might ask what needed to be named, in this period, that did not require naming earlier. The word seems to take its poignancy from a tension between the acceptance of hierarchy, on the one hand, and the claims of equality, on the other. This tension was produced, and thus became something worth mentioning, at the moment that the balance between the values of hierarchy and equality, respectively, began to shift (for a heterogeneous set of reasons). As that shift of balance continued, and the value of equality came to trump that of hierarchy, the seventeenth-century sense of the word became less plausible and eventually fell out of use. *Condescension* came to take the meaning it has today. Not only did the semantic range change, but so too did its coordination with other words, such as *insolence*, as well as the gestures, body language, clothing, financial strategies, and housing arrangements (recall Prospero's back rooms) that work with them.

Condescension and the other terms that worked in coordination with it came into being as nameable things—they were objectified—when certain aspects of social interaction seemed worth noting and making explicit, and they fell out of use, in their original senses, when the social world no longer supported them. But the word *condescension* itself persisted, now functioning to name a new idea, the insult to the taken-for-granted value of equality when people act in a patronizing way, as if they know they are superior to another.

This historical process of coming into view and falling back into obscurity again is a version of what has been called "problematization." This is a term introduced by Michel Foucault in the context of writing about the relationship between freedom and ethics. Like many historians and social scientists, he wrestles with the implications of social determinism for people's emancipatory projects, or what in the context of this chapter we might call the possibilities of moral progress. In Foucault's case, this leads to a question that, as we have already seen, lies at the heart of virtue ethics: How can we make sense of an individual's efforts at self-formation in a world of naturalistic and socioeconomic causalities? On the one hand, although Foucault rejects in principle any appeal to a given "human nature," he also dismisses the idea that people are simply free to invent themselves in any old way. But he also rejects the claim associated with the "science of unfreedom" noted in the introduction, that historical forces so overwhelmingly shape human possibilities that freedom is an illusion.

Foucault's answer to the problem of determinism draws on a long-standing position in the Western tradition of moral thinking. He assumes that to be ethical is to have some freedom of action and, so, to discover the ethical is to reveal some fundamental distinction between causal determinations of behavior (whether those be biological, psychological, or socioeconomic in origin) and people's capacities to act. With echoes of Adam Smith (1976), Foucault proposes that the freedom in question derives from the fundamental human capacity for reflection. Reflection, in turn, depends on self-distancing:

> Thought is . . . what allows one to step back from [a certain] way of acting or reacting, to present it to oneself as an object of thought and to question it as to its meaning, its conditions, and its goals. Thought is freedom in relation to what one does, the motion by which one detaches oneself from it, establishes it as an object, and reflects on it as a problem. (1997: 117)

If ethics depends on the freedom that is made possible by reflection, then the fundamental cognitive capacity for self-distancing discussed in chapter 1, and developed through the self–other dynamics reviewed in chapters 2 to 4, is an ethical affordance. If self-distancing turns the flow of action into an *object* of thought, then this cognitive process of objectification, the third-person perspective it can produce, and the semiotic means that facilitate and sustain it are devices for ethical life—but only in some modalities: for as those earlier chapters argue, ethical life is confined neither to those emotions, intuitive responses, and habits that elude consciousness nor to reflexive self-awareness.

One very ordinary kind of objectification is naming. This refers to the creation of verbally explicit categories and descriptions and their application to specific persons and actions. When a writer notes the duke's act of "gracious condescension," any reader can appreciate that something more than describing lunch is at stake: an ethical judgment is being offered. That judgment says something about the duke's character and the writer's stance toward hierarchy. The reflexivity induced by ethical problems works with some fundamental properties of human cognition, such as the capacity for metarepresentation (Bloch 2012; Sperber 2000). Making habits or intuitions the objects of reflection can give them the cognitive status of facts, for example, about one's own character or about moral truths, or they can take the form of rules. In some cases, it seems that this process can give them a high degree of stability and accessibility. They can enter into the settled "descriptions" that are presupposed by what the philosophers call "action under a description" or the social scientists call "the definition of the situation." Objectification, for example, in verbal formulas ("right to life," "woman's choice"), can help support what cognitive scientists call "crystallized attitudes" (Bassili 1989; Schwartz and Bohner 2001).[2] These are relatively fixed judgments or positions (abortion is wrong, a woman should control her own body) that require relatively little cognitive processing, and are less susceptible to the bias imposed by framing effects (or at least

[2] I thank Fred Conrad for helpful discussion of this topic.

new ones), because they have been worked out in advance and are readily available to awareness.

Granted that self-distancing and the third-person perspective are basic human capacities, what induces them to come into play in any given instance? What might prompt one to "step back" and question oneself? What makes that easier or harder to do? We have already seen part of the answer in the previous chapters: certain moments of social interaction fail to flow smoothly, leading to a loss of the sense of shared reality, or for some other reason require people to give one another reasons, justifications, excuses, accusations, or some other explicit account of what is going on or what kind of persons they are. But where do these accounts come from, and why, even within a single community, are they persuasive sometimes and not others? Sometimes people are brought to awareness of some contradiction between competing values within a single social world. Sometimes the change arises because coexisting values come to be juxtaposed in new ways, making their incongruities apparent. According to Foucault, the classical Greek ethics of pleasure was rendered problematic when confronted with that of self-mastery (Faubion 2001: 98). In nineteenth-century England, the value of deference to hierarchy was problematized by the growing emphasis on equality. Or the pressure may be exerted by social conflict, for instance, when members of a less powerful group begin to demand rights from a more powerful one (Anderson 2013).[3] As we will see, objectifications and the reflexivity they facilitate can play a key role in catalyzing the changes that give ethical life a history.

[3]I want to forestall the temptation some readers might have to assume that the difference between habitual and objectified ethics maps onto one between so-called traditional and nontraditional societies. On the one hand, even suburban American society is full of habitual ethics. And conversely, even the more "traditional" members of mid-twentieth-century Navajo society could be ethically highly articulate. Their insistence that people must not rely on compulsion when making decisions together meant that they had to rely on persuasion. As a result, it has long been commonplace in Navajo decision making to talk things out, giving everyone involved a say, with the result that people were highly practiced in giving reasons for their positions (Ladd 1957: 203–6). In Navajo life, it seems, ethical problematization is constantly stimulated by the egalitarian values that inform their styles of interaction.

ABOLITIONISM

In the previous chapters we saw a variety of ways in which ethics can become a problem that prompts reflexivity. These range from conversational repair to scenes of instruction (such as Zinacantan gossip or a Los Angeles father's "teachable moment") and interactions in which one party can demand an account from another (as in excuse making or judicial hearings). But problems like these do not necessarily produce ethical change; they might remain isolated moments of individual self-awareness, or they might simply end up reenforcing familiar norms.

One example we looked at, however, clearly catches historical transformation on the fly. The conflicting voices of Don Gabriel, the farmer caught between the values of Mexicano villagers and those of the encroaching world of Spanish speakers, arose from the challenge that each ethical world posed to the other. It is not merely the case that two worlds collide and that eventually, one will triumph over the other. More than that, the clash induces new forms of self-awareness, as people such as Don Gabriel are forced to articulate values and practices that had once been taken for granted. The position of the Spanish speaker gave him an outsider's perspective on himself as a Mexicano villager. This is one version of the distancing that makes reflection possible. The different voices he expressed are steps in the direction of full-fledged objectifications. True, when we encounter Don Gabriel, his voices remain somewhat subliminal, matters of style, prosody, intonation, and vocabulary. But one can imagine a point at which villagers will be deploying explicit ethical categories ("venerable traditions," "entrepreneurial vigor") and taking guidance from them in their actions (for instance, rejecting the obligations of kinship in favor of the freedom to take advantage of new commercial opportunities and being able to justify oneself in doing so).

In Don Gabriel's clashing voices we have a glimpse into one of the major forces driving ethical history. Social conflict is a powerful source of problematization. But some have also argued that social conflict is crucial to moral progress as such. One of the most convincing cases for such progress is the abolition of slavery, especially in cases, such as that of Great Britain, where the story is not clouded by the multidimensional regional conflicts that run through the American case. Looking at the abolition of slavery, the political philosopher Elizabeth Anderson (2013, forthcoming) offers one compelling version of how social conflict can, in her view, contribute to ethical progress. Nineteenth-century abolitionism, like the temperance and early women's movements, has been called a moral crusade by scholars of social movements (Calhoun 2012: 264). It could not be fully explained within the usual terms of material or political self-interest. Whereas three hundred years ago, the legitimacy of slavery throughout much of the world was rarely questioned by those who might have been in a position to do so, today, even though

> *de facto* slavery persists in many areas of the world, virtually no one is willing to defend it. Slavery stands today as a paradigmatic moral wrong. The transformation of moral consciousness that underwrote the abolition of slavery represents perhaps the most profound instance of moral progress the world has ever seen. (Anderson 2013: 2)

How did this happen? The British abolitionist movement arose and triumphed with astonishing speed. The first Quaker petition against the slave trade came out in 1783; Parliament abolished the trade just twenty-four years later, before eliminating slavery in the colonies in 1833. But this is not a simple story of moral suasion. After all, the founder of Quakerism, George Fox, had been preaching against slavery in 1671, on the grounds that all humans are equal in the eyes of God (notice the technique of self-distancing that invoking the perspective of God entails, something to which we return in the next chapter). More than a century passed before even the Quakers were induced to act. So if self-interest is not sufficient to explain the movement, neither is the moral case against slavery. Nor are the fundamental human propensities for empathy and the sense of fairness. Once invoked, they would provide emotional power and even a sense of inevitability to the abolitionist cause—but slavery's long history makes it clear that they were not sufficient to prompt opposition to it in the first place.

The historian Christopher Leslie Brown points out that the arguments against slavery were widely accepted in Britain long before anything was done about it. He shows how many contingent factors had to come into play before those arguments could give rise to actions, including the patriotic claim that Britain was the world's bastion of liberty, an equally patriotic hostility to the Americans, and the emergence of new social identities among women, religious nonconformists, and workers. It was a "moment when various individuals and groups found through their challenge to the Atlantic slave trade an opportunity to establish new identities, new self-conceptions, to create for themselves a new place within society and a new role in public life" (Brown 2006: 2). Evangelical enthusiasm and popular nationalism both imparted new energy to the cause:

> Often activists took up the issue of slavery less because they cared about Africans than because they regretted its impact on society, on the empire, on public morals, or on the collective sense of self. . . . [S]elf-concerned, self-regarding, even self-validating impulse in early British abolitionism represents a key theme in this book. (Brown 2006: 25)

Ethical history, in this light, depends on factors so heterogeneous and contingent that it is hard to imagine that it can be explained just by the acceptance of increasingly abstract principles, expanding circles, or any other overarching pattern. But Anderson draws from the same events a more general proposition: changes in ethical awareness can be prompted by social contention, making claims against more powerful groups by or on behalf of less powerful ones. Contentious actions can include moral arguments, petitions, publicity campaigns, litigation, boycotts, demonstrations, and riots. Anderson proposes that if contention is successful, it will do three things: inform the powerful of the interests of others less powerful than them, challenge their sense of legitimacy, and display the moral authority of those who challenge them. What Anderson is describing is one version of that process of taking a distance on oneself—being forced to see oneself through the eyes of others—which can produce ethical reflection. But as she argues, this distance-taking must be situated in a context of practical action, not just contemplation, if it is to effect real changes in ethical awareness.

Ethical awareness, in this account, follows not just from changed thoughts but from changes in the circumstances that sustain them and make them practicable. For instance, as slavery loses its legitimacy, and people begin to stop supporting it, the peer pressures to hold others to its norms, and conform to them oneself, will slacken. The dynamic is social (and draws on the psychology of norm-seeking described in chapter 1), not just conceptual. But Anderson (2013: 10) maintains that reflection is needed to prompt people to seek out alternative ways of living. Once a new way of living becomes fully established in laws, institutions, economic arrangements, and so forth, then the absence of slavery seems merely normal. At that point, subsequent generations will find it impossible even to imagine that slavery could be legitimate or that their own moral emotions and sense of self-respect would ever permit them to accept it if it were.

CONSCIOUSNESS-RAISING

If we grant that all sorts of contingent factors contribute to ethical transformations, what, then, *is* the place of reflection, explicit talk, and argumentation in ethical history? Obviously they are not accurate representations of people's motives or goals or the emotional or cognitive processes these involve. But that does not mean we should dismiss them as simply misleading or false (pace Bloch 2012). As objectifications they mediate social interactions. Their public availability can afford new developments. For example, explicit ideas are more readily subject to criticism and revision or rejection than more tacit assumptions, precisely because they are more visible. As Anderson suggests, one result is that explicit ethical talk can help create or deflate the legitimacy of an existing norm and foster the search for alternatives. This is because explicit concepts are more directly subject to people's awareness—a general feature of human cognition, not a particular claim about certain social arrangements or cultural constructions.

As the story of abolitionism suggests, changes in ethical life depend on social conditions that will sustain them. This is why it can be hard to distinguish between ethics and politics: in the slogan of mid-twentieth-century American feminism, "The personal is political" (see Hanisch 2006). By treating abolitionism or feminism as part of ethical history, I do not intend to reduce politics to ethics. But there are dimensions of political life that cannot be understood without some grasp of the moral and ethical impulses behind them. As we have seen, this is especially clear in the case of activists whose political commitments cannot be directly explained in terms of their own self-interests, such as bourgeois intellectuals who fight for the proletariat (Karl Marx, Friedrich Engels), men who defend women's rights (John Stuart Mill), or literati who aspire to emancipate peasants (Ho Chi Minh)—or an ordinary social worker such as my neighbor Sally, who was introduced at the start of this book. Moreover, ethical transformation can be an important outcome of political movements, even if those movements fail in their immediate goals. As we will observe in the chapters that follow, focusing on ethics can help us see more clearly the parallels and links between social movements that identify themselves as political and others, such as evangelical religions, that do not.

At this point, I want to look at a brief moment in the American feminist movement when the ethical effects of awareness played a central role. This is the invention and promulgation of the technique of "consciousness-raising." Consciousness-raising is interesting for several reasons: it took very seriously the effects of problematizing the habits of everyday life, it succeeded in changing the descriptions and evaluations of actions and persons that were available for many Americans, and it ultimately foundered, in part, on an unresolved tension between subjective experience and objective social analysis. It gave rise to new terms such as *sexism*, *sexual harassment*, *date rape*, *marital rape*, and *eating disorder*. As I write, American court cases are reformulating the moral language of rights, justice, freedom, equality, and happiness in the context of gay marriage. It seems that in doing so, they are following the lead of some changes in public attitudes that are occurring with breathtaking speed. These changes are influenced by, among other things, a long-standing strain of libertarianism in American thought, the ideals of companionate marriage, and recent sociological effects of "coming out," as people discover that well-known relatives, colleagues, neighbors, and celebrities are gay. As these factors become established, it seems that in many cases gay relationships are being drawn into a sphere of ethical concern where nongay people will discover that their basic capacities for empathy and the sense of fairness apply. Clearly, however, the moral emotions are not the driving force behind this change. After all, as with slavery, historically most Americans did not feel that empathy and fairness applied to same-sex relationships; instead other capacities equally foundational to ethical life such as disgust and third-party norm enforcement did. But, as with slavery, should the new ethical ideas and institutions become sufficiently well established, they will, in effect, recruit those capacities for empathy and the sense of fairness. We may speculate that this recruitment would eventually foster ethical sensibilities for which the new social arrangements are both emotionally compelling and conceptually a matter of mere common sense.

The changing status of same-sex relationships is part of a longer history in the ongoing transformation in the ethical understanding of sexuality and gender in the United States and Europe. The current wave of change draws, at least in part, on transformations in ethical awareness that were set in motion by an earlier generation, even if the immediate political aims of those feminists were not fully realized. If we go back to the moment in which the slogan "The personal is political" was coined, we find a highly purposeful attempt to render everyday habitual life explicit in order to make it a knowable object and therefore something that could be grasped and transformed. This was the consciousness-raising group.

As with abolitionism, feminism drew on a range of contingent factors. Consciousness-raising was introduced to a women's liberation convention in 1969 by a group called the Redstockings. Like many radical feminists of this period, the young women who made up this and similar groups had come through the civil rights movement and the New Left but had become disenchanted by their marginality within male-dominated organizations (Freeman 1973; Rosenthal 1984; Sarachild 1978). Despite this, these women still "lived in an atmosphere of questioning, confrontation, and change. They absorbed an ideology of 'freedom' and 'liberation'" (Freeman 1973: 805). Along with ideology, they brought with them tactics and extensive networks that became crucial to the rapid growth of the feminist movement after 1967.

The rise and fall of the politically radical version of consciousness-raising was rapid. Publicized by *Time* magazine in 1969, the method spread rapidly, while its political message grew more muted (Buechler 1990: 121; Hanisch 2006: 2, 2010; Rosenthal 1984: 317–24). By the mid-1970s it had become part of the equipment of the general American therapeutic culture. The narrative as told from the perspective of political activists is often one of disappointment. Rather than affecting mass mobilization, it seemed to result in individual personal self-realization. But viewing it as a moment in ethical history, it may be that consciousness-raising had deep and long-lasting effects in middle-class American life. It was, after all, a kind of problematization that forced reflexivity on large areas of daily life that had gone largely unexamined before. The full impact of consciousness-raising in American life is a story for the historian to investigate. Our concern here is to take a closer look at what happens when taken-for-granted experiences give rise to objectified categories of ethical judgment.

FROM PERSONAL EXPERIENCES TO ANALYTICAL CATEGORIES

Consciousness-raising was meant to form a bridge between change in individuals, on the one hand, and large-scale social transformation, on the other. With this in mind, the idea of "consciousness" purposely conflated ideas with two distinct origins. According to one early proponent, "'consciousness-raising' . . . refers to a long and logical process which leads to a *synthesis* of the personal consciousness to which the psychoanalysts have given their attention, and the political or class consciousness of the Marxists" (O'Connor 1969). The technique drew on the "rap sessions" developed in the civil rights movement, published accounts of "speak bitterness, recall pains" meetings in communist China, and New Left theories, notably those of Herbert Marcuse, whose emancipatory arguments stressed the primacy of consciousness in social change within a Marxist framework (Rosenthal 1984: 312).

What these diverse sources had in common was an emphasis on the socially transformative role of individual awareness and the effectiveness of relatively abstract categories of analysis in making sense of concrete personal experiences. But the process started in the reverse direction, from the personal to the social. The underlying assumption was a kind of empiricist version of Marxist theories of class consciousness: people living under similar material conditions will have similar subjective experiences, but they do not initially realize this. Lacking the concepts that would reveal their similarities to one another, women think that their difficulties are the result of personal failures and inadequacies. It is only once individuals compare experiences that they will discover what they have in common. Generalizing from this, they will then be able to create more abstract categories, such as patriarchy or sexism, which will enable them to connect individual sources of unhappiness to social conditions of oppression. Thus the general categories that emerge from particular experiences are brought to bear back onto experience, allowing one to see particular events as instances of general types.

Because this technique was based on the idea that new categories of analysis would emerge from previously unexamined experience, it meant, at least in principle, that participants should reject prior sources of authority in favor of the first-person perspective: "For every scientific study we quote, the opposition can find their scientific studies to quote.... Everything we have to know, have to prove, we can get from the realities of our own lives" (Sarachild 1978). But the process would not stop at the first person, since what should emerge from the comparison of personal experiences is a common thread, as women find "that in fact they have all been telling the same story with minor variations" (O'Connor 1969). The group is

> a place where the members see their experiences mirrored in each other, where they are able to check and reaffirm their perceptions. One woman alone who complains of her oppression can be told she is distorting reality. When it happens enough she learns to doubt her own observations.... But when a group of women perceive again and again the same patterns of oppression derived from concrete stories of their day-to-day lives, it is impossible to sweep away their words as

> distortions. The first stage ends with a collective recognition that their tales of failures and feelings of inferiority are not functions of inferior people, but of some unnameable force that has acted upon them all to make them feel inadequate. (O'Connor 1969)

The process of recognition should lead to the naming of that hitherto "unnameable force."

Although the participants seemed to assume that the categories could be discovered directly in their experiences, inevitably they drew on the existing resources of social thought (see Scott 1991). Thus, in what seems to have been the moment in which the technique itself was first dubbed, a woman named Ann Forer is reported to have said, "I've only begun thinking about women as an oppressed group and each day, I'm still learning more about it—my consciousness gets higher" (Sarachild 1978). That is, having acquired the category of "oppressed group" in contexts where the expression referred to African Americans or to the proletariat, she found that it could apply as well to women. One result, according to research carried out with early gay rights consciousness-raising groups, was the development of new group identities, new "minorities," that the participants could apply to themselves (Chesebro, Cragan, and McCullough 1973: 136). Much like the looping effects described by Hacking, once these categories began to circulate, people could find new ways to recognize themselves in them. They became new ways of being a person and of entering into social relations with others.

Yet perhaps even more transformative than the categories of analysis were the process of working them out and the raw materials on which they were brought to bear. Consciousness-raising sessions drew on the ordinary structures of conversational interaction and intersubjectivity. They depended on the ways in which self and other affirm and recognize one another and, when things are going well, develop and protect a shared sense of reality together. Earlier I quoted Jack Sidnell's observation that it is an ethical question whether one is being attended to by others. The consciousness-raising group responded to this dimension of interaction and pushed it further. As Kathie Sarachild put it, "One of the exhilarating and consciousness-raising discoveries of the Women's Lieration [*sic*] Movement has been how much insight and understanding can come from simple honesty and the pooling of experience in a room full of women" (1978). In many cases, it would seem that the ethics of the interaction was at least as powerful as any particular final analysis or practical outcome.

As for the raw materials, part of the power of consciousness-raising—as well as a source of vulnerability both to ridicule by outsiders and to worries by some political radicals—lay in the attention it drew to the apparently trivial details of ordinary lives. Talk turned to the dieting, wearing of uncomfortable clothes, and playing dumb that women undertook for husbands and bosses:

> In a small group it is possible for us to take turns bringing questions to the meeting (like, Which do/did you prefer, a girl or a boy baby or no children, and why? What happens to your relationship if your man makes more money than you? Less than you?)." (Hanisch 2006: 3)

As Sarachild writes, "We hadn't realized that just studying this subject and naming the problem and problems would be a radical action in itself" (1978). It was a process of rendering the habitual and taken-for-granted available for inspection and critique.

Although the purpose was to stimulate political action, the process began with an ethical transformation through the redescription of everyday life. As the moral philosopher Alison Jaggar puts it,

> Simply describing ourselves as angry, for instance, presupposes that we view ourselves as having been wronged, victimized by the violation of some social norm. Thus, we absorb the standards and values of our society in the very process of learning the language of emotion, and those standards and values are built into the foundation of our emotional constitution. (1989: 159)

The politics of that anger depended on the prior existence of some ethical sense. The ethical dimension of consciousness-raising is similarly apparent in Sarachild's remark that the discussions touched on "areas of the deepest humiliation for all women" (1978), for whatever else humiliation is, it is a feature of the ethics of interaction. It arises from the self-evaluations that people draw from their interactions with one another. In one account, a woman described how men at a political meeting had ignored her suggestions, which left her feeling that

> everything she had to say was stupid and trite, and furthermore because she was too ugly and unpopular to be noticed. She found it hard to tell her story to the group because she believed it reflected and revealed some horrible private personality characteristics—stupidity, ugliness, sickness, and dependency on other's approval which she interpreted as emotional flabbiness. (O'Connor 1969)

Finding a common thread in these stories helped participants shift the locus of responsibility away from themselves: "The most important is getting rid of self-blame. Can you imagine what would happen if women, blacks, and workers . . . would stop blaming ourselves for our sad situations?" (Hanisch 2006: 4). Recall the trial of the man who murdered his son, mentioned above. In that case, Charlson was exonerated by virtue of a causal explanation from neurophysiology: the court concluded that he was not at fault for his actions, because of the effects of a brain tumor. In consciousness-raising we see a parallel form of exculpation, drawing on ideas about sociopolitical causality. Their success depends in part on women taking up a third-person perspective on their own lives, seeing events as instances of more general processes. Like humiliation, blame is an ethical category, one that depends on assumptions about how people account for themselves as agents. As we will see in the next two chapters, both religious and communist movements aimed to a similar relocation of agency in people's ways of evaluating their worlds. First, however, let us take a closer look at how the transformation of personal experience into objectified categories can work.

RECONSTRUCTING ETHICAL FEELINGS

The feminist philosopher Naomi Scheman analyzes one effect of feminist consciousness-raising groups, the discovery by individual participants that they had been "angry" for years, without realizing it:

> To discover what we are feeling (our emotions) is not necessarily or usually to discover some new feelings . . . ; rather, it is to discover what all of that means, how it fits in with who we are and what we are up to. It is to put a name to a mass of rather disparate stuff, to situate

> the otherwise rather inchoate "inner" in a social world, to join (introspectible) feeling and behavior in a significant way, to note a meaningful pattern. (1980: 174)

Writing of a hypothetical participant she calls Alice, Scheman says that

> she became gradually more aware of those times when she felt depressed, or pressured and harried, as though her time were not her own. However, she didn't believe her time ought to be her own, so in addition she felt guilty. . . . She didn't think she had any *reason* to feel this way; she never took the bad feelings as justified or reasonable; she didn't identify with them; they came over her and needed to be overcome. (1980: 176–77)

In her consciousness-raising group, Alice is encouraged to acknowledge and express those feelings, and she finds that they are treated as legitimate and justifiable by others.

These feelings preexist the act of naming and identifying them as a certain emotion: the stimulated amygdala, the surge of adrenalin, and so forth are not social constructions. What changes, rather, is their cohesiveness and their social standing. As suggested in chapter 3, in order to function as a moral emotion, the state of arousal we might call "anger" must be understandable as anger about something, a condition the arousal itself is not sufficient to determine. In Scheman's story, the members of the group urge Alice to see her anger as justifiable. To say that anger is justified and legitimate is not necessarily to discover a new emotion or even to alter an old one (although Schachter and Singer's experiment with epinephrine described in chapter 1 suggests that such changes may indeed occur). Rather, in Scheman's view, it is to redescribe the situation so that the emotion becomes a legitimate response. Scheman treats this redescription as fundamentally political. But it is also a reconfiguration of the ethical world, introducing new conceptual objects in the form of new types of persons (e.g., male chauvinist pigs, liberated women) and new act descriptions (e.g., sexual harassment). In the process of deciding that her feelings are justified, they become part of Alice's self-understanding, as a person of a certain sort and certain values. This is much like the self-cultivation of virtue ethics, and, as we will see in the next chapters, the process of conversion to a religious faith or a revolutionary worldview.

But the crucial changes are not simply transpiring within the interior life of an individual. First, of course, they emerge through social interaction itself, in the consciousness-raising group. Beyond that, in redescribing the situation, what is being changed is "the picture we are likely to have of what the good life for a woman consists in" (Scheman 1980: 178). The anger becomes legitimate because the woman's current way of living turns out not to be the only one available. New possibilities, new ways of being a woman—what virtue ethics would call new kinds of human flourishing—have emerged. By making these possibilities explicit, these sessions made them available for people to recognize in other people and other circumstances and to respond with criticism, enthusiasm, hostility, support, or just plain confusion.

More than that, the group produced forms of self-awareness that depended on other people: "A frequently remarked feature of such groups is that each woman's ability to recognize and change her situation depends on the others doing the same" (Scheman 1980: 180). It would seem that this was for two reasons. The Alice of Scheman's example recognizes that others have had the same experiences; they transcend any particular subjectivity. She also sees that the others recognize her as someone whose feelings are justified. These forms of objectified self-understanding affect interactions down the road. Once others know that Alice is a feminist, they have certain expectations for how she will feel and behave. As with the reputed murderer in Zinacantan or the piously veiled Kabyle woman, mentioned in previous chapters, those expectations will shape other people's responses to her, which in turn will influence her own possible actions and interactions. This public context—consisting of other individuals and the categories and expectations they bring to bear on one another—can provide Alice with some of the "exoskeleton for character" the helps sustain an ethical stance.

Scheman is not denying that feelings have a real psychological or physiological basis. The question, rather, is what they amount to: "What makes the whole affair a case of someone's being angry is how it is as a whole similar, in ways we particularly care about, to other cases" (Scheman 1980: 183). The criterion for including a particular instance of feelings under the rubric of "anger," in this view, is not its resemblance to other feelings in the abstract (provoked, say, by an incident of road rage, a botched plumbing job, seeing a flag-burner, facing an armed enemy in battle) so much as it is its resemblance to other situations of sexism. In other words, what makes "women's anger" identifiable as such is the fact that "we [feminists] particularly care about" each case of it. The identification of anger with the social category of women depends on a political analysis, but the way feminists care about it depends as well on an ethical vision of the good life. The politics is inseparable from the idea that a way of flourishing has been denied. It is on the basis of that sense that one's flourishing has been denied or thwarted that the feminist can link one distinct situation to another, treating both as instances of a single category: "anger."

As the history of abolitionism suggests, feminist anger cannot change ethical life all by itself. The social history to which it contributes would include such things as the forming of activist organizations, changed employment and family laws, the opening of formerly male universities to women, the advent of easy birth control, shifts in religious authority, the creation of academic programs in women's studies, and vast numbers of books and films. But the newly objectified descriptions of formerly unself-conscious habitual daily life surely helped catalyze these actions and animate their supporters.

The various experiences of women that came to be identified under the rubric of "anger" emerged through processes of social interaction that problematized aspects of daily life that had been largely unexamined—O'Connor's unnameable force. The outcome was a new self-description tied to an ethical stance, albeit one that was linked to a political argument about injustice and oppression. It became possible to say, "I am angry (because that's not the way a good life should be lived)." Consciousness-raising groups could not do all this alone, but they seem to have played a key role in having what the philosopher Hilary Putnam (1975) called a "baptismal effect," giving a name to something so that it could be identified—links forged to other things (domestic labor, for example), a crystallized lexicon that could be applied across instances (that one woman might discover that she has the same anger that another one just expressed)—and granting social authority to those who can identify it. The process was called consciousness-raising because its proponents recognized that the fact of giving something an explicit verbal formulation, rendering it cognitively available, could have consequences for people's relationships and actions well beyond the moment of the group's own meeting. To the extent that the social movement that gave rise to these groups sought ethical revolution, it did so, in part, by demanding a shift from habit to awareness. Feminist consciousness-raising may have been short-lived—its effects sometimes more individualist than political in scope and its relevance less apparent to those women who were also grappling with problems of race, poverty, or sexual orientation—but the method of inducing reflexivity through interaction has been used in movements for ethical reform and revolution in all sorts of historical contexts. Among these are religious movements. These have been the preeminent cases of successful, long-lasting, highly cohesive morality systems. This is the topic of the next chapter.

CHAPTER 6

Making Morality in Religion

Ethical Life and Morality Systems

The previous chapter approached ethical history by focusing on processes that make ethical concepts and categories the explicit objects of people's awareness. The formulation of previously unrecognized ethical concepts—one goal of consciousness-raising—might remind us of the story in the introduction, about the Polish peasant who remarked that a Jewish girl was "not a dog, after all." This woman's act—striking, of course, for its rarity—gives us a microcosmic if dramatic case of objectification: the girl has been identified with a knowable, nameable category in ways that draw out certain expectations from the ethical background in ways that impose them on the awareness of others. But even in less dangerous times, such momentary interventions are likely to remain evanescent. What makes an ethical innovation stable and socially viable, repeatable in some recognizable way across different contexts, accepted within a larger community, perhaps overriding competing ethical stances?

As the examples in the previous chapters suggest, language plays an important role. But ethical cultivation and education may also work with gesture (Lempert 2012), song (Harkness 2014), agricultural activities (Pandian 2009), hunting prohibitions (Valeri 2000), reproductive technologies (Roberts 2012), grocery shopping (Miller 2001), legal activism (Dave 2012), accounting techniques (Miyazaki 2013; Poovey 1998), market bargaining (Keane 2008), and the handling of money (Akin and Robbins 1999; Hart 2001; Mandel and Humphrey 2002; Parry and Bloch 1989; Zelizer 2005). Ethical life is sustained by patterns of child rearing and adult peer pressure, sermons and stories, laws and working conditions, kinship systems and medical practices, and so forth.

In any instance, however, these various practices do not necessarily add up to a coherent and consistent whole. What Bernard Williams calls a morality system depends on the coordination of what might otherwise be disparate ethical ideas and practices. That coordination is not something to be taken for granted. Yet, because morality systems are typically easy to see—they announce themselves through their rituals, disciplines, rules, texts, authorities, slogans, laws, and justifications—they loom large in the historical and ethnographic scholarship. Their visibility makes it easy to forget that nothing guarantees that any given social world will produce a coordinated and explicit morality system or, if it does, that the resulting morality system will actually govern people's ethical lives in their entirety. At the same time, if we react against this bias toward morality systems and insist that ethics is really and only a matter of the unself-conscious habitual practices of everyday life, it becomes hard to account for the empirical existence of morality systems when we *do* encounter them. Moreover, it becomes hard to understand certain kinds of actions, people's purposeful efforts to change ethics. For morality systems loom large not just because they are easy to see but also because they play such a large role in history. Morality systems are often shaped by self-conscious people who aspire to stand apart from the taken-for-granted flow of life in order to act upon it.

This chapter and the next one examine two different kinds of morality system, religious in this chapter, self-consciously atheist in the next. What these morality systems share is a propulsive movement, as large numbers of people take action in order to transform their ethical worlds. They are expressions of historical agency. Given the evidence we have seen in previous chapters, that much of ethical life is instinctual, habitual, fragmentary, and contradictory and does not require anyone's full awareness, how do we make sense of the morality systems that push hard against all these characteristics? What even makes it possible to step outside the flow of life and look at it from a critical distance? In this chapter we will try to understand the demand for moral consistency that motivates a pair of thriving contemporary piety movements, one Christian, the other Muslim. Although they differ in their theological and moral doctrines, these movements have much in common. In particular, the participants in these movements actively and self-consciously strive to live ethically consistent lives. In both piety movements, that demand for consistency is partly explained by the inculcation of a God's-eye view, a version of the third-person perspective from which the faithful is expected to see the totality of his or her life and impose order on it. But not all morality systems depend on God for that perspective, and chapter 7 turns to the Vietnamese communist revolution, in which a Marxist theory of history played a similar totalizing role.

HISTORICAL OBJECTS

According to Williams, a morality system has a juridical character and implies the existence of a godlike judge that transcends the plane of human activity. Ethics, by contrast, involves the growth or cultivation of persons, guided by models or concepts of human flourishing that vary according to the social and historical context. If the quintessential morality system is a monastic order, the paradigmatic ethics is Athenian virtue. Yet some ethical projects of self-cultivation, such as those of Tibetan monks (Lempert 2012) or Jain world renouncers (Laidlaw 1995), do take morality systems as their model for human flourishing. Such morality systems are often marked by a high degree of objectification, producing descriptions of acts (for example, lists of sins) and types of persons (for example, saints) that are readily cognizable. They are like the descriptions that emerge when one must give an account of oneself, as discussed in chapter 4. Morality systems support awareness with a host of semiotic technologies, such as texts, pedagogic techniques, and rituals, and institutions to sustain them, such as law courts, temples, and schools. These produce what we can call *historical objects*. They are objects because they have an explicit character that people can focus on cognitively, much as one might learn a set of rules. They are historical both because they show persistence and stability and because they are subject to transformations. Rules work in part because they seem to be the same even when they appear in different contexts. One reason they are subject to transformation is that, being explicit objects of awareness, they are easier to criticize and defend than the more subtle patterns of unself-conscious habitual life. This openness to criticism and defense makes ethics available for purposeful endeavors such as religious revivals, social movements, and political revolutions.

Historical objects are words or practices that make certain aspects of ethical life cognizable because they have semiotic forms that, being durable, can be recognized again and again. Words such as *sexism*, practices such as veiling, and bodily habits such as bowing have a degree of solidity in people's experience and can easily draw attention. Like descriptions of kinds of persons (honest, sneaky) and kinds of actions (generous, cowardly) whose reappearance in new settings can lend a sense of coherence to the flow of events, they are part of the raw material out of which everyday ethical awareness is made. This sense of fixity and palpability sets them somewhat apart from the background of those psychological processes and patterns of interaction described in chapters 1 to 4. An important part of ethical history is the way in which aspects of ethical life emerge out of this background and become historical objects.

Semiotic mediation means that even inner thought, if it is to have any social consequences, is never just cognitive but is inseparable from a host of material practices. Those practices in turn are embedded within historically specific ways of life. Having books or a system of schools or a monastic system or military orders or yoga poses will inevitably produce different results for ethical life than not having them. However, those consequences are not defined in advance. The existence of historical objects and the technologies and institutions that sustain them does not determine any particular ethical outcome: they afford reflections that may (or may not) catalyze change. Notice the feedback loop that enters into these processes. When people produce historical objects, they are likely to be responding to something that psychology or the pattern of social interaction affords. Once those objects become part of the background for particular communities, they in turn offer affordances: ritual practices aiming to influence gods may become psychotherapy for atheists, martial arts meant to hone fighters might become meditation regimes for literati, universities intended to train clerics for church and court may become homes to laboratory scientists and nationalist historians.

TAKING ETHICS IN HAND

The remainder of this chapter sketches out some approaches to these questions: Assuming that what we call "religion" and "ethics" are in principle distinct from each other, what is the conceptual relationship between them? What are the historical pathways along which the two often seem to converge? What are the social implications of that convergence where it occurs? And when they converge, what remainder escapes the conflation of these two? Our focus is on people's purposeful efforts at pious self-fashioning as analyzed in studies of two religious revival movements at the end of the twentieth century, the charismatic Christianity of the Urapmin of Papua New Guinea (Robbins 2004) and the self-cultivation of Muslim men in Cairo (Hirschkind 2006).

The Urapmin (whom we met in chapter 3) are a small group in the highlands of Papua New Guinea. Never missionized, they actively set out to convert themselves starting in the 1960s, sending young men and women to study with Baptist missionaries in neighboring communities. This set the stage for the enthusiastic reception in 1977 of a revival movement that was sweeping the region. The entire community accepted the religion, emphatically rejecting anything they associated with "the ways of the ancestors." Since this moment, they have been energetically working to construct a completely Christian life for themselves, aiming at the radical transformation of both individual and the entire community. To that end they engage in frequent prayers, group spirit possessions, and confessions and take part in church services in which they are subjected to lengthy harangues about their lack of self-control.

Robbins describes their condition in the 1990s as one of "moral torment." The torment in question arose from contradictions between the older ethics embedded within daily activities and forms of sociality, on the one hand, and the explicit precepts of the new religion, on the other. For despite their rejection of the old religious practices, Urapmin still hunt, work their gardens, and enter into social relations much as before. Social relations depend on established forms of recognition, especially gifts and cooperation. To be a good person, in this social world, involves balancing two sets of demands. On the one hand, one should restrain one's willfulness in order to live up to the model of generosity, reciprocity, and helpfulness expected within established relationships (Robbins 2004: 195). On the other hand, one should also work to create new relationships. To do this, however, means exercising one's will, which is likely to be at the expense of existing relationships. For example, instead of giving meat to an affine where it is due, one might give it to someone with whom one is developing a new connection. If the tension between morality and the value of new relations already existed in pre-Christian society, it is brought to the foreground in the constant Christian attack on willfulness.

According to Robbins (2004: xxiv), Urapmin think of morality in terms of lawlike dictates. Their dilemma is thus implicit in the tension between the taken-for-granted domain of everyday practices and the highly self-conscious realm of doctrinal teaching brought by Christianity. But the problem is even deeper. Their version of Christian morality defines it in opposition to any this-worldly calculations of utility. This came to the fore when, under the influence of Western apocalyptic ideas, they abandoned their fields in expectation of the incipient end of the world. The very impossibility (Robbins 2004: 246) of perfectly fulfilling the demands of their version of Christianity is part of what defines a rule as being moral in the first place. This induces the sense that, despite their best efforts, they are ultimately doomed to live as sinners as they persist in the ordinary, this-worldly business of sustenance and communal life.

Consider now another religious revival movement, the subject of Charles Hirschkind's research with Muslim men in Cairo's lower-middle- and lower-class neighborhoods in the 1990s. Although this dense urban setting could not be more different from Papua New Guinea forests, like the Urapmin, these Cairenes are deeply influenced by religious currents that have been sweeping the globe, in this case the Islamic revival. They too have made the religious movement their own and actively work to achieve profound moral transformations, both personal and social. This activist relationship to moral change takes place against an everyday background of modern urban commerce and religious institutions. Theirs is a community in which many (if not all) aim to make religious morality a dimension of everything, not just a specialized domain set apart from the rest of life. And like Urapmin, these men in Cairo find ethical life to be difficult and to require constant, highly self-conscious effort. But, nevertheless, it remains an attainable goal, which can and should be integrated into daily life. One of the distinctive practices of these men is habitually listening to sermons on cassette tapes. As Hirschkind puts it,

> Played in public transport, in shops, garages, and cafes, sermon tapes reconfigure the urban sound space, imbuing it with an aural unconscious from which ethical reasons and action draw sustenance. . . . Sermon tapes enable their listeners to orient themselves within the modern city *as a space of moral action* . . . . [C]assette-sermon audition is also a technique of self-fashioning. (2006: 22)

What is crucial to the movement is the insistence that ethics is a dimension of practical activity, not something that stands in contrast to it or something the cultivation of which requires withdrawal into a pious enclave kept pure by its barriers against an unpious world, unlike, say, communities of Hasidim (Fader 2009) or Russian Old Believers (Rogers 2009). Whatever torment may haunt these men is found not in contradictions within ethical life, or even exactly between that and their worldly context—for any difficulties the latter most pose them should in principle be surmountable—but in the threat of hell for the unfaithful. That threat is represented in graphic terms within the sermons themselves. If these men aspire to ground their ethical lives in automatic bodily responses to religious summons, highly explicit reminders of the afterlife constantly prod them to see themselves from the third-person perspective, provided in this case by the God's-eye view.

Both accounts, Robbins's of Urapmin and Hirschkind's of Cairo, portray ethics in terms of *both* deontology and virtue. That is, on the one hand, each piety movement posits explicit sets of rules, understood as relatively abstract principles that have very general application; and, on the other hand, in practice each movement also fosters the inculcation of certain concrete habits and specific ways of living guided by a vision of the good person.

ETHICS AS PIETY

For all their differences, these two works portray movements that share a basic starting assumption, that religion is the preeminent domain within which ethical reasoning and action occur and the culmination of ethics is piety. Urapmin Charismatics and Cairene Muslims both treat theology as the necessary authorization and sufficient justification for ethical actions. Religious institutions and practices are the chief practical means by which the ethical life is shaped. Now I should be clear, both studies focus on highly energetic, purposeful religious reform and purification, or piety, movements, which are not typical either of the broad range of long-established Christianities or of Islam as lived even by all Cairenes. However, in one respect, the Urapmin and the sermon listeners in Cairo are exemplary of an influential strand of popular thinking in many parts of the contemporary world. This is the assumption that religion is the necessary foundation for ethics. The assumption that ethics requires a religious basis seems to be an important factor (if, no doubt, only one of several) driving the present global religious revival, whose effects are manifested in both Urapmin millenarianism and the Islamic revival movement in Cairo. It is an assumption held even by many secular observers of contemporary society who are not themselves speaking from a religious position, such as the sociologists who write:

> If, today, religion ceases to be a resource for public reflection and nothing replaces it, our values and morals lose their foundation. That loss is clearly evident in the contemporary discourse about human rights. . . . Everyone now agrees that human beings have such rights. In contrast, no one can say from whence these human rights derive or on what they are founded. (Porpora et al. 2013: 3–4)

We will return to the question of human rights in the conclusion. What I want to stress here is that all the evidence of the previous chapters should show why ethical life does *not* necessarily require religious foundations. When people come to think that ethics *does* depend on religion, this is for historically specific reasons.

HABITUAL ETHICS

Hirschkind's account of the pious men in Cairo draws on Foucault's (1985, 1997) adoption of classical Greek ideas about self-fashioning, which emphasizes the role of mentors, role models, and other supportive resources in the individual's community. This view of ethics focuses on habits and practices embedded in the ordinary activities of everyday life (Lambek 2010). Often those habits and practices are considered to be most valuable when they are so deeply part of the person as to lie below or beyond consciousness. Conscious or not, virtue is not confined to particular domains of life; any given aspect of social existence potentially bears ethical weight. This portrayal of virtue as habitual and pervading communal life has many of the hallmarks of what the anthropologists who follow Émile Durkheim call a "total social fact." One of the best examples of the ethics of the total social fact is Marcel Mauss's description of "the gift." In societies such as the precolonial Polynesian Maori, the Northwest Coast Haida, or the Norse, where social relations were traditionally negotiated through the formal exchange of valuables, the gift involved

> an enormous complex of facts. . . . Everything intermingles in them, everything constituting the strictly social life of societies that have preceded our own. . . . In these "total" social phenomena, . . . all kinds of institutions are given expression at one and the same time—religious, juridical, and moral, which relate to both politics and the family. (Mauss 1990: 3)

Mauss places ethical intuitions and expectations at the center of the everyday practices of law, economics, and kinship. The concept of the total social fact thus pushes back against the tendency to privilege religion as the basis of ethical life. Although the modern division of labor attempts to sort out domains such as the religious, the juridical, the domestic, the economic, and the political, this is a historically specific, and ultimately (according to Mauss) not entirely successful, endeavor. But of course history matters, and this sorting out of domains is enormously consequential (Keane 2007: 83–91; Smith

1962). What happens to the concepts and practices of ethics when they come to be identified with religion?

In the rest of this chapter I will sketch the following argument: scriptural monotheisms tend to objectify ethics, exerting pressure on them to become more consistent and cognitively explicit. But objectification also tends to separate ethics from everyday habits and foster the taking up of a third-person perspective on ethical life. This tendency is reinforced by the modern division among what the sociologist Max Weber (1978) called "value spheres." Paradoxically, this very division of value spheres, by identifying ethics with a specifically religious domain, can inspire strenuous efforts to break out of that confinement. The result can be the cultivation of piety meant to encompass *all* domains of life. Yet the piety project must still wrestle with the consequences of that initial objectification, which tend to undermine that very goal.

THE GOD'S-EYE POINT OF VIEW

Both the Urapmin and Cairene piety movements are monotheistic and scripture-based. These movements presuppose a category of "religion" that is already the outcome of a history of sharpening distinctions among different spheres of action. Religion in these movements emerges in *contrast* to the ubiquity and taken-for-granted everydayness that characterize the social fact. Many contemporary versions of the idea that ethics depends on religion derive from this distinction among value spheres. But some scholars of present-day capitalist societies have shown that value spheres are not so sharply distinguished, revealing ethical assumptions implicit even in practices that seem to be defined by purely means–ends rationality, such as those in business and politics. This is why even in bastions of free enterprise, we still see moral repugnance at the idea of buying and selling sex, or children, which economic models might treat as wholly irrational (Zelizer 2005; cf. Anderson 1993; Sandel 2012). This repugnance need not have a specifically religious basis: it can be part of the largely unself-conscious ethical intuitions embedded within the everyday life of a community.

By contrast, the scriptural, monotheistic religions develop a point of view located *not* in the immediate context, such as the clan or the first-person position of the rational maximizer, but, rather, in the person of a single supreme being. Whether they stress divine transcendence or immanence, monotheisms characteristically posit a unifying perspective that demands of ethical life that it have a high degree of principled organization. The monotheistic perspective is expressed in the following remark about the effect of Islam on tribal Arabia. According to the religious scholar Toshihiko Izutsu, Islam marked

> the first appearance of moral principle which was consistent enough to deserve the name of "principle." A whole practical code of conduct, though as yet largely unsystematic, was imposed upon the believer.... [Prior to Islam, moral values] were just there as membra disjecta, without any definite underlying principle to support them.... Islam made it possible for the first time for the Arabs to judge and evaluate all human conduct with reference to a theoretically justifiable moral principle. (2002: 196)

When Izutsu sees this prior condition merely as the *absence* of principles, he is taking on the totalizing perspective of the monotheistic moral code. But the state of being that Izutsu calls "membra disjecta" is hardly some failing peculiar to early Arabia. Rather, the ethnographic record shows that it is a familiar state of affairs, in the absence of some centralizing ethical project.

The explicit goal of totalizing ethics under a theoretically justifiable moral principle, in religion, stands in contrast to the totality implicit in the Maussian social fact. The latter refers to the ways ordinary social existence is already thoroughly saturated with ethics prior to any regulating principles. As the earlier chapters tried to show, social interaction is saturated with ethical matters of values, obligations, the demand for recognition, and the risk of its denial. These do not require any overarching organizing principle in order to have their effects. As the Chinese philosophers of effortless action (*wu-wei*) recognized, social interactions are usually carried off most fluidly, most successfully, by people who do not have to monitor themselves constantly. This does not mean that they can be understood without reference to evaluation, judgment, and self-awareness—only that these do not need to be organized into a single coherent system in order to have ethical effects. That is, even a true gentleman who always knows the right thing to say and do is probably acting on a case-by-case basis, and not mentally referring to a system of etiquette in which all the parts fit together. As I argued in chapter 4, some ethical worlds accord a privileged role to self-consciousness, and others, to habit, but nothing in the concept of ethical life makes this a matter of either/or, confining it wholly to one basis or the other.

One distinctive project of any monotheistic religion (albeit not one confined to monotheisms, as evident in Buddhism, Confucianism, and Jainism) seems to be the effort to rationalize ethics under an organizing principle. Or at least, the positing of a transcendental point of view is likely to instigate such an effort. This is because such a point of view invites a universalization that seems to demand principles sufficiently general that they can hold across an indefinite number of cases and contexts. These seem to be preconditions for both global proselytization and ethical generality. The point of view of a transcendent deity offers a position on which to stand, from which one may survey the whole range of known ethical values available in any given cultural world, such that their inconsistencies become visible. It is the pressure exerted by this asymptotically transcendental point of view that provides at least the conceptual and ethical motivation for the kinds of purification or reform movements that are so characteristic of monotheistic religious history.

One way to universalize and purify an ethical principle is to strip away particularities. The cost, however, can be a kind of debilitating dematerialization and undermining of social relations. Robbins says that prior to their conversion, Urapmin possessed

> techniques of self-formation (observance, sacrifice) that allowed those who employed them to achieve the telos of living as "good people" whose wills were under appropriate control. Christianity, in proclaiming their codes false, also condemns their techniques of self-formation. In doing so, it leaves the Urapmin without any way to constitute themselves as ethical subjects in their relations with the spirits or things of this world. They are left, instead, with a Christian ethical system that counsels them to look to their inner lives as the only target of their ethical efforts. (2004: 223–24)

This is not to say that there aren't alternative practices available in monotheistic traditions, but the transcendentalizing move always contains the latent possibility—the invitation—of further purification, such as iconoclasm, antiritualism, and other attacks on the material things and practices that prior religions had made use of.

ENTEXTUALIZATION AND SACRED TRUTH

Unlike clan-specific taboos, ritual regulations, and the like, a universal abstract ethical principle is not likely to render itself immediately inhabitable. Some further specification and mediation is needed. Perhaps the most ubiquitous medium for rendering abstract theology accessible is entextualization. *Entextualization* refers to the processes by which specific chunks of discourse are rendered into texts, by eliminating or altering linguistic features that ground them in a specific context such as first- and second-person pronouns (Silverstein and Urban 1996). As texts, they are transportable away from particular contexts in order to circulate among a potentially indefinite range of other contexts, where they have the potential to be recontextualized (Keane 2004, 2007). They are thus, among other things, crucial elements in the portability of any religion that aspires to universality and engages in proselytization.

More generally, the process of entextualization is, virtually by definition, what makes a so-called scriptural religion. At the conceptual core of the scriptural religions are the products of entextualization, texts extracted from one context that can have powerful effects when recontextualized in another: scriptures, creeds, catechisms, liturgies, sermons, prayers, hymns, and so forth. The practices surrounding these texts—studying scripture, reciting creeds, learning catechisms—are also crucial to their realization in adherents' lived experience. To be sure, those practices coexist with meditation, pilgrimage, sacrifice, trance, dance, charitable works, and so forth, but it is often the texts that mark the point of distinction between more and less legitimate religions (possession of scripture is, for instance, central to the legal definition of "religion" in Indonesia).

These texts are often highly explicit. This explicitness reflects the way in which the scriptural or doctrinal text is available as a reply to some kind of demand that one provide foundations or reasons for ethical claims. To be sure, many religious texts are obscure, even opaque, and all of them require interpretive strategies if the reader is to engage with them. But they are discursive in their very nature: they represent ethics in words, a process that already involves a degree of distance from embodied practice and of generalizing beyond particular circumstances.

The Qur'an is a good example. The multistage transmission from revelation by the Archangel Gabriel to the speech of Muhammad to its inscription by scribes at his dictation thematizes the entextualization process. The authority of the text that results cannot be wholly detached from this chain of transmission. The ethically authorizing nature of the scriptural text is clear:

> It is one of the most characteristic features of Qur'anic thought that it conceives of "religion" in terms of the "guidance" of God. In this conception, the religion in the sense of *islām-imān* is nothing other than *ihtidā'* (verb, *ihtadá*), which literally means "to be rightly guided" or "acceptance of guidance." This is but a corollary of the basic fact that, in the Qur'ān, Revelation is regarded as essentially a merciful guidance (*hudá*) for those who are apt to believe. Indeed, even the casual reader of the Qur'ān would not fail to notice that through the whole of it there runs the fundamental thought that "God guides whom He will," or—which would, logically, collide with the preceding—that God is absolutely fair in giving guidance graciously to all men, but some people accept it while others reject it of their own free will. In either case, the revealed "signs" are divine Guidance. (Izutsu 2002: 193)

I want to point to three things about this passage, which represents at least one line of orthodox thinking on the subject (see Rahman 1994: xvi). First is the emphasis on direction. Scripture may offer all sorts of things—a theodicy, an eschatology, an ontology, mythical exemplars, models for poetic language, compelling parables, liturgical texts, political legitimation, a script for performances—but what really matters here is its role as a medium for conveying divine guidance for the ethical life, practical advice for how to live. Second is the strong sense of exteriority: ethical guidance by this account comes to people from something beyond them, which they can accept or reject. Indeed, this separateness (and its intuitive appeal to the moral psychology of third-party norm enforcement) is a condition of its ethical character to the extent not just that ethics depends on adherence to a law or ritual procedure but that it is a *chosen* adherence, since one could have rejected the law (that is, although "submission" is definitive of Islam, it must in principle result from an act of free will). The transcendental nature of this divine point of view, which posits a single organizing vision on what might otherwise be a fragmented field of ethical norms and moral injunctions (Izutsu's "membra disjecta"), is implicated in its separateness from humans.

Certainly this is only one, albeit highly orthodox, version of Islam, and it is speaking within a theological discourse rather than responding to the particular needs and expectations of specific formations and problems of domestic life, politics, law, art, commerce, and so forth. But it shows how Islam connects ethical behavior to knowledge of the truth about reality. Ethics hinges here on accepting truth, which is revealed in signs (*ayat*). As Rahman puts it, "In order to determine the meaning of a sign, one must have, in addition to reason, a certain disposition, i.e., the capacity for faith" (1994: 70). Immorality is, in this light, tantamount to willful disbelief, which involves the denial of the signs of the Creator. Thus the factual assertions of the creed are foundational to the ethical demands of the moral system.

This way of linking assertions of belief to ethics forms a marked contrast to the more or less tacit everyday ethics embedded in social interactions we have seen in the previous chapters. These do not require any particular set of explicit truth claims for their moral authority. To be sure, underlying concepts can often be elicited by an analyst, for instance, that the giver is coextensive with the gift or that all humans are created free and equal. But nothing about everyday ethical life requires that such beliefs be asserted—or even available to anyone's consciousness. Indeed, as James Laidlaw puts it, "Some ethical traditions at least are only liveable *insofar* as they may be apprehended as fragmented narratives whose relation to each other is left somewhat unclear, thus enabling people to shift as best they may between them" (2014: 82).

This is one reason why notions of sincerity tend to be so much more prominent in creed-based religions (see Keane 2007), since the acceptance of truth claims is so closely linked to ethical judgments. It is this reference to truth claims, and the textual organization of those truth claims, that helps give the sense of coherence, of being a matter of context-independent principles, to scripture-based ethical systems. The demand that ethics be based on a small number of general principles reflects this link to knowable truth and its roots in the transcendental perspective from which the alternatives seem merely incoherent, so many "membra disjecta."

ABSTRACTION AND STRUGGLE

For all their differences, both the Charismatic Urapmin and pious Cairenes take ethics to depend on heightened awareness, through both doctrinal knowledge and self-monitoring. This awareness is reinforced by the tensions and conflicts between piety and the unmarked habits of daily life. For even the goal of rendering piety a matter of deep, instinctual everydayness seems to require the pious to maintain constant alertness to the impious possibilities around them. In both the Urapmin and Cairene cases this tends to result in ongoing struggle. This is due, in part, to how religion has become a distinct category within a world of secular institutions, mundane activities, and less pious contemporaries.

The more the faithful identify ethics with piety, the more unethical those less pious people and practices, among whom they find themselves every day, are likely to seem. Yet, unless they follow the route to separatism, those who aspire to piety cannot wholly abandon every one of those habits and social relations that tie them to other persons and practices, at least not without social difficulty and personal loss. This difficulty is not just a practical matter of making a living and sustaining bonds of kinship, neighborliness, or workplace fellowship. Beyond those practical matters, those everyday habits and relations are likely to possess their own ethical implications, whose pull may not simply be eliminated by the ethics of piety. Piety notwithstanding, in everyday life one is still likely to acknowledge that a decent person owes something to kin and colleague. Referring to other research in contemporary Egypt (Schielke 2009), Laidlaw comments that

> sincere followers of reformist Islam also think and feel in moral registers other than those of piety. They care about and aspire in terms of social justice, community and family obligations (respect), good character, romance and love, and self-realization. These can all be in conflict in various ways. (2014: 171–72)

Paradoxically, the totalizing pressure toward the full integration of pious personhood appears to stem, in part, from the very distinction between religion and nonreligion it seeks to overcome. These persistent tensions seem to be an important source of the restless urgency of those revival movements that identify ethics with piety.

The discussion of piety in this chapter dwells on the conceptual and imaginative features of self-consciously religious movements. In the next chapter we turn to a communist revolution that tried to inculcate a secular equivalent of the God's-eye point of view in order (in part) to broaden the moral circle. The revolutionaries tried to teach people to see themselves in objective social and ethical categories. As in consciousness-raising sessions and religious confessions, their efforts built on the affordances of everyday interactions and their capacity to induce reflexive objectifications. And as with the other reforms and revivals we have looked at, the practical preconditions and unintended consequences of reform make it impossible to tell a simple story of ethical history.

CHAPTER 7

Making Morality in Political Revolution

THE ETHICAL ATTACK ON RELIGION

The previous chapter examined the familiar assumption that religion is a necessary foundation for morality. This assumption is widespread—evident in everything from early Buddhism to contemporary Islam—but it is historically contingent and cannot fully account for either the sources or the goals of ethical life. Certain religious ethical regimes do, however, possess two features that make them especially important for understanding ethical life as a sociohistorical phenomenon. One is their explicitness and systematicity. Most of the globally dominant religions known today make morals a matter of awareness and aspire to overall consistency. The second is their universalism, often linked to a sense of historical inevitability: they aspire in principle, if not always in practice, to push the moral circle ever wider. As we have seen, both features make use of the God's-eye view. But do universalism and the expanded moral circle *require* religion? Can there be a God's-eye view without God? Are the reasons religions give to justify ethical commitments necessarily more compelling than others? In this chapter, we will consider the case of moral revolutions that are bound up with political ones that are self-consciously atheist.

One of the hallmarks of the twentieth-century socialist and communist revolutions was the effort to remake societies that were more or less dominated by religious faith on nonreligious or even militantly antireligious grounds. Marxist revolutions shared an assumption held by many modern critics of religion, that it undermines the true basis for ethical life. The idea has various sources. One is the anticlerical notion, shared with some religious reformers, that religious institutions deflect people's ethical regard onto themselves and away from its proper objects (of course Marxists often saw churches as political competitors too). But another has to do with self-awareness. Socialist revolutions tended to develop their own versions of a view of the moral narrative of modernity that had wide purchase by the end of the nineteenth century. As noted in chapter 5, this narrative depicted the gradual emancipation of people from their belief in false gods, onto whom they had displaced the powers that are properly their own, which could now be reclaimed (Feuerbach 1957; Marx 1978). As a result, for instance, Soviet regimes sent professional teachers of atheism into the countryside to debunk religion, teach elementary science, and enlighten the peasants (Greene 2010; Husband 2000; Luehrmann 2011; Peris 1998). The purpose was not just general education but ethical transformation. Once villagers learned not to expect miraculous interventions by God and His saints, or to look to priests for moral guidance, they would become responsible for their own actions (Keane 2014).

Obviously these revolutions had political and economic goals. But they cannot be understood without this ethical dimension. Ethical objectives were a necessary, if not a sufficient, motive force behind revolutions (especially for activists who were not themselves direct victims of oppression, like many of the intellectuals who helped instigate peasant and worker movements). As the historian E. P. Thompson (1971) and political scientist James C. Scott (1976) argued long ago, even oppression is not a purely objective condition. Reflecting on peasant uprisings in Southeast Asia, Scott remarks that "woven into the tissue of peasant behavior, . . . whether in normal local routines or in the violence of an uprising, is the structure of a shared moral universe, a common notion of what is just" (1976: 167). Both ethical motives and goals of ethical transformation ran through the twentieth-century socialist and communist movements, not just in Russia but in Africa (Donham 1999; McGovern 2013), China (Kleinman et al. 2011; Madsen 1984; Oxenfeld 2010; Stafford 2013), and Latin American (Lancaster 1994; Theidon 2013; Wilson 2014). These revolutionary movements did not rely on the redistribution of economic resources, land reforms, and new forms of governance alone.

To say this, of course, is not to reduce the political and economic dimensions of revolution to the ethical, only to insist that the ethical cannot, for its part, be reduced to the political or economic.

Given that any large-scale political movement is necessarily shaped by a complex and contingent mix of motives, goals, strategies, and tactics, there are no straightforward stories to tell about revolution. But this chapter will consider one in which the idea of revolutionary ethics played an important and relatively durable role at both the level of leadership and among many more ordinary participants. It focuses on some of the ethical sources and goals of Vietnam's anticolonial and communist revolution. By looking at how, in everyday practices, revolutions attempted to propagate an expanded moral sensibility, inculcate people with egalitarian values, and reconfigure their intuitions about agency and responsibility, we can see some of the links among psychology, face-to-face interaction, and social history.

ETHICAL SOURCES OF VIETNAMESE REVOLUTIONARY THOUGHT

The historian William J. Duiker concludes his authoritative narrative of the Vietnamese revolution by asking what explains the success of the communists over their rivals. After considering such factors as their superior organizational abilities, their skills in military and political strategy, the talents and charisma of Ho Chi Minh, and the vacuum left by a historically weak middle class, he concludes that a crucial element was their higher moral credibility for many ordinary Vietnamese. He (1995: 255–56) endorses the argument of the Marxist intellectual Nguyen Khac Vien, who held that the older Confucian emphasis on collective responsibility, personal ethics, and the responsibilities of elites predisposed many intellectuals to Marxism in the early years of the movement. Like neo-Confucianism, Marxism is this-worldly, altruistic, and convinced that the truth lies in texts.[1] That affinity was part of the leaders' appeal to merchants, students, and, ultimately, rural villagers.

[1] Victor Lieberman, personal communication, April 15, 2014.

The leaders understood this, for the revolution was, among other things, a movement of ethical reform. Ho Chi Minh famously said, "If you want to build socialism, before all else you must have socialist people" (quoted in Malarney 2002: 56). What this meant, in part, was the inculcation of "revolutionary ethics"; as he also remarked, "If you do not have ethics, you do not have a foundation" (quoted in Malarney 2002: 53). This reversal of orthodox Marxist theories of revolution, in which material conditions bring about changes in consciousness, is evidence of the extent to which Ho remained rooted in older ethical doctrines of Vietnamese elites. Modifying traditional Confucian lists of the virtues, he identified the five cardinal virtues of revolutionary ethics as benevolence, righteousness, knowledge, courage, and incorruptibility (Malarney 2002: 54). By all accounts, he was quite earnest about this and made every attempt to model the virtues by his own example. According to the historian David Marr, "Ho Chi Minh . . . consistently tried to project certain exemplary virtues, such as simplicity, punctuality, physical fitness, cleanliness, direct speech, and personal intimacy . . . [which] served as models for several generations of cadres" (1984: 98–99).

And yet, of course, Ho was also a master of political analysis and ruthless strategy. Vietnamese communists captured the apparatus of government, defended it for decades against French and American assault, expanded their control into a resistant south and ventured into Laos and Cambodia, undertook major land redistributions by force, engaged in a prolonged attack on those defined as oppressor classes, and introduced ideological indoctrination throughout the countryside. But running through the welter of indoctrination and mobilization, class conflict, anticolonial yearnings, economic reforms, military violence, and the sometimes harsh tools of governance were the threads of an attempted moral revolution.

Vietnam's communist movement arose out of a heterogeneous mix of other anticolonial and religious forces that were contending for supremacy in the first decades of the twentieth century. Vietnamese who resisted French domination or otherwise sought to reform or revolutionize society were influenced by neo-Confucianism, French secular socialism, conservative French moral pedagogy, Buddhism, Catholicism, the new religion Cao Dai, and a number of millenarian and charismatic sects (Tai 1983). Vietnamese Marxism, of course, had roots in an engagement with the wider spectrum of European thought and morality systems. Ho Chi Minh was educated in French colonial schools; leaving Vietnam for Europe on the cusp of adulthood, he did not return until the age of fifty-one. His anticolonialism and socialism, however much they may have been provoked by the immediacy of Vietnamese experiences with colonial oppression, drew on French thought and later the institutional structures of the French Communist Party and the Comintern (Marr 1971: 276)—although his sense of entitlement as a moral leader was rooted in Confucianism. French books, speeches, and organizations gave manifest form to the articulate historical objects of egalitarian and leftist thought. The theoretical texts and the redescriptions of the world they offered seem to have played a role similar to that of scripture in the piety movements discussed in the previous chapter. That is, from the distant perspective afforded by Marxist historiography and class analysis, Ho was in a position—a secular version of the God's-eye view—that enabled him to see the Vietnamese situation in a new light, as a comprehensive whole, with new standards against which to evaluate it and with new sets of categories allowing it to be compared with situations elsewhere. After the communist triumph in northern Vietnam, one of the pressing tasks was to bring these categories into the everyday life of people who had not been so educated, leveraging high theory in order to induce the ethical transformation of everyday life.

At the start of the twentieth century, the ethical thought of Vietnamese elites was dominated by neo-Confucianism. It centered on the cultivation of virtue, which is expressed in the "five relationships": between ruler and subject, father and son, husband and wife, elder brother and younger brother, and friend and friend. All but the last of these is inherently unequal. Within this set of assumptions, it is worth noting what counted as "moral." For one thing, morality was not merely a set of rules to be followed, a set of external laws and institutions (*phap*). Rather, the inner quality of virtue (*duc*) was conveyed by the example laid out by the higher member in each pair. Confucian ethics, in this context, bears some resemblance to Aristotle's, namely, the downplaying of rule-following, the role of exemplification, the place of face-to-face relations, and a dependence on social (and therefore moral) hierarchy. Another distinctive feature of morality in this tradition was self-denial. In this respect, the Confucian view of ethics, as understood in modern Vietnam, bore one of the hallmarks of a morality system. Even the lowliest member of society could achieve virtue by forgoing personal gain in favor of collective obligations (Marr 1984: 58). That is, the sense of obligation was made evident in the suppressing of self-centered instrumental reasoning. One of the challenges the revolutionaries faced was to transfigure these concepts while retaining their moral force. In particular, this involved maintaining the sense of obligation without the sustaining structures of hierarchy and holding onto the demand for self-denial while vastly expanding its moral purpose to encompass a far more abstract object, not kin, clan, or village but party, nation, and class.

EVERYDAY ETHICS, EVERYDAY OPPRESSION

By the beginning of the twentieth century, the effects of French colonial rule were felt in different ways across the social spectrum. For the political classes, loss of political autonomy and humiliation of the court undermined Confucian confidence in the cosmic connection between ruler and moral order. The concerns of peasants might not have been so theoretical or idealistic. They voiced an earthy skepticism with maxims such as "Win and you're called Emperor, lose and you're a bandit" (Marr 1984: 103). Living under highly constrained circumstances, villagers tended to make ethical judgments in terms of a moral economy trimmed according to what they took to be reasonable and just possibilities. In contrast to both the literati and the villagers was the small class of shopkeepers, managers, civil servants, students, and others in the nascent technical or professional fields. Aware of both French norms and Vietnamese traditions, and in frequent personal contact with the colonizers, they were especially vulnerable to everyday humiliations. But the denial of social recognition occurred across the social spectrum:

> The French, from their positions of prestige, routinely engaged in petty acts of racial discrimination. They infantilized their male house servants

> by calling them "boys." They frequently referred to young women as "*congaie*"—from the Vietnamese "*con gái*"—which simply means "girl" in Vietnamese but which the French commonly used in a pejorative sense to refer to their concubines. French settlers belittled Vietnamese in numerous other ways, such as by the use of the personal pronoun "*tu*" generally used with intimates and inferiors. (McHale 2004: 43)

According to Marr, the high colonial era led Vietnamese thinkers to a preoccupation with ethical questions. As Confucianism became increasingly discredited, Vietnamese themselves began to write instructional texts influenced by French moral pedagogy. By the 1920s and 1930s, these proliferating texts were becoming "obsessed with the nature of Vietnamese customs and their transformation" (McHale 2004: 75). Radical political thought only slowly gained traction during the decades before World War II; it was morality that most concerned the literate Vietnamese, as is evident in the large numbers of essays and letters to the press on this subject they were publishing (Marr 1984: 55).

Yet the outcome of these moral concerns could feed back into the development of revolutionary politics. As elsewhere in Southeast Asia, the first decades of the century were what a historian of Java has called "an age in motion" (Shiraishi 1990). Literate people in the cities were enthralled with the new forms of subjectivity and emotional vocabulary they could glimpse through reading, schools, interactions with Europeans, and travel. In these circles,

> a fascinating debate was opened on the meaning of love (*tinh-yeu*, *ai-tinh*) in its various manifestations, which was connected with a complete reassessment of the relationship of the individual to his family and to society. . . . Out of the emotional self-searching, the moodiness and despair, came something that one perceptive author has characterized as revolutionary romanticism . . . that would drive many thousands of young men and women to devote their lives to renewing the struggle against the colonialist. One youth, having left home for a life of revolutionary danger, summed up this drive in a simple parallel sentence which was forwarded to his mother's funeral: "At a time of agony for our countrymen, we have taken Vietnam as our mother; for mourning garb we will emulate Mazzini." (Marr 1971:

> 264, quoting Dan Thai Mai, *Van Tho Cach Mang Viet-Nam Dau The Ky XX* [Hanoi, n.d.], 143; footnote numbers removed and internal quotation reformatted)

This situation encouraged ethical "problematization" (see chapter 5). What was new was not social conflict or moral conflict, which are enduring features of social life, but, rather, novel contexts in which people like the young man quoted above were compelled to account for themselves to one another, making their ethical choices explicit. This man could justify his apparent defiance of filial piety by substituting nation for mother, allowing him to retain his commitment to an ethics of obligation. Vietnamese could also draw on a growing range of conceptual sources allowing more distant perspectives on the taken-for-granted habits of ethical life and the local experiences of everyday life. By 1945, the revolutionaries' declaration of independence could easily invoke both Jefferson and the French Declaration of the Rights of Man and Citizen.

REVOLUTIONARY ETHICS

After gaining control in the north, the communists actively pursued programs to construct socialism's new ethical person. They construed this person in terms drawn from Confucianism and other sources, reconfigured in light of Marxist modernity. During the war against the French (1945–54), a training manual for cadres portrayed the classic virtue of "humaneness" (*nhan*) as the highest revolutionary virtue, but in an altered form, maintaining that

> the virtue of humaneness consists of loving deeply and of wholeheartedly assisting one's comrades and compatriots. That is why the cadre who displays this virtue wages a resolute struggle against all those who would harm the Party and people. That is why he will not hesitate to be the first to endure hardship and the last to enjoy happiness. That is why he will not covet wealth and honor, nor fear hardship and suffering, nor be afraid to fight those in power. Those who want nothing are afraid of nothing and will always succeed in doing the right thing. (*Let's Change Our Methods of Work*, quoted in Marr 1984: 99)

When Communist Party cadres entered the rural districts, they were guided by several assumptions about the contrast between revolutionary and peasant ethics. First, in the communists' view, Confucianism had treated peasants as the passive targets of a moral indoctrination. In keeping with the idea that socialism requires the transformation of ethical persons, the party took seriously the role played by the conceptual categories and social types of Marxist theory. Categories such as "society," "citizen," and "public opinion" had been deployed by urban intelligentsia of various political stripes since the 1920s (McHale 2004: 33). What was new was the earnest attempt to bring them to uneducated rural people. As they took control in the north after World War II, communist cadres set about educating the villagers in the new descriptions of their world—what I've called new historical objects—and the means of applying them to concrete instances.

Although the concepts were derived from elite social thought, often abstract and difficult for the uninitiated to grasp, much of their power for villagers came from their links to moral judgment. Through regular indoctrination, public denunciation sessions, and techniques of self-criticism modeled on Maoist practices of "speaking bitterness," the cadres propagated new definitions and evaluations, such as wicked exploiting classes, and induced villagers to apply them to specific individuals (Luong 1992: 186–94; Malarney 2002: 23ff.; McHale 2004: 141). In later years the lessons were reinforced by pervasive posters, banners, slogans, poems, songs, and even job application materials (Leshkowich 2014), which came to be part of people's experiences and memories, much like, it would seem, the well-remembered television commercials of their contemporaries in the United States. And yet the process was not entirely one of top-down didacticism. Given the contrast to the passivity supposedly induced by traditional ethics, "commandism" was to be avoided, and people were to be persuaded (Malarney 2002: 68). This attention to persuasion reflected problems of mass mobilization that many revolutionary and reformist movements have faced (Cody 2013). One difficulty was simply keeping people motivated to struggle for distant goals. Another arose from the paradox that villagers had to be induced to become self-motivated agents. As a result, by the time revolutionary reforms took root in the village, some had been altered, and others, dropped, as communists sought workable solutions at those points where villagers most resisted them. The emphasis on persuasion also meant that reasons had to be given, and policies, justified—often by appeal as much to ethical values as to practical needs of the moment.

The second way in which communists distinguished themselves from tradition was the ethical devaluation of material self-interest. Marxism holds that individuals' material interests and their shared relations to the means of production produce the class identities driving class struggle. But, paradoxically, the revolutionary effort depended on idealistic individuals' self-denial in the name of a greater good beyond the self. In Marr's view, "from a moral viewpoint, at least, it was probably the peculiar Marxist combination of materialist self-interest and idealistic claims regarding the perfection of human character that kept many Vietnamese involved." He continues,

> It cannot be denied that those Vietnamese Marxist-Leninists who chose immortality and perfection of mankind as a whole over salvation of the individual in the immediate world were rated as exemplary revolutionaries.... In this sense, exemplary individuals made the moral decision to deny themselves present gains in favor of rewards for posterity.... [T]he pressure on cadres to lead spotless personal lives and be totally dedicated to the cause was considerable. (1984: 129–30)

It is hard to miss the resemblance to the pious radicals discussed in the previous chapter. Ethical revolution was both an ultimate goal and, for many individuals, a considerable motivating force in the first place.

The communists' third point of distinction was an emphasis on dynamism and progress that was set in opposition to what they took to be peasant fatalism. This was closely tied to the attack on religion and promotion of science. According to a primer published in 1968,

> When the masses still believe in the heavens, spirits, and fate, they will be powerless before natural changes and the difficulties that the old society has left behind. People will not be able to completely be their own masters, the masters of society, nor the masters of the world around them. (Ninh Binh, *The Task of Building the New Ways, New*

Person and the Progressive Family in the Struggle against America to Rescue the Nation, translated and quoted in Malarney 2002: 80)

Having imbibed Victorian social evolutionism with their Marxism, revolutionary leaders assumed that spirit worship was a survival from a primitive stage when people lacked the ability to control nature (Taylor 2004: 38). It followed from this moral narrative of modernity that only after the conceptual liberation from fetishes, gods, and other illusions would humans be restored to the practical agency that is properly their own and thus take ethical responsibility for their circumstances. In the Marxist version, whereas religion looks to distant agents and past practices, the revolutionary vision is future-oriented. The realization of that future depends on the responsibility of humans to exert the agency they now realize is theirs. To become emancipated in this way is a crucial part of what it means to be modern and why modernity has distinctly ethical implications, and not just technological, economic, or sociological ones.

The process of emancipation required that villagers adjust their understanding in both social and ontological terms. Socially, they had to be persuaded that landlords, not fate, were the causes of their suffering; ontologically, they had to replace the entire religious worldview with that of science so that, better understanding real causality, people could take control over their lives. Secularization campaigns in the villages therefore attacked anything deemed superstitious, from the lunar calendar, which marked out auspicious and inauspicious days, to amulets, votive offerings, and divination. What defined the category of superstition was belief in supernatural agents or causalities, deemed immoral because they supplant human responsibility.

Finally, the fourth way in which communism was distinguished from tradition was in the effort to expand villagers' moral circle. The narrowness of the traditional circle of moral concern was captured in a village saying, "No one mourns for the father of everyone" (Malarney 2002: 51). The traditional sense of moral obligation was focused on the person's immediate kin, especially parents and ancestors. Beyond kinship ties were relations of spatial proximity. Shaun Malarney (2002: 14), who carried out fieldwork in the northern village of Thinh Liet in the 1990s, observes that residents took their primary field of moral reference to be their own village, judging one another by how well they embodied the values of equality, solidarity, reciprocity, and affective relations identified with village life. Thus the range of ethical concern was tightly focused on those neighbors with whom one was most likely to enter into everyday face-to-face interactions, which were always closely observed for signs of respect or disrespect. That ethical concern was manifested more publicly in wedding and funeral rites, where it could be measured in one's demeanor and material contributions. Certainly the efforts of the Communist Party expanded the moral circle; yet Malarney's finding indicates that the preeminent domain of immediate ethical concern remained centered on the scale of face-to-face interaction, where deference and recognition were most immediately registered.

The expansion of the circle of moral concern, combined with the promotion of modern notions of agency and responsibility, converged in the reformers' attack on village ritual life. The rituals of ancestor veneration were a target of concern for several reasons. First, people propitiated ancestors in part to ward off supernatural punishment, a displacement of human ethical responsibility due to unscientific concepts of causality. Second, by focusing ethical attention solely on one's own kin, ancestor rituals worked against the expanded moral circle according to which one should identify with larger social orders such as the nation, one's class, the party, and so forth. The greatest ethical resistance to communist attacks on religion concerned precisely those acts that fulfill the moral obligations of filial piety. The prototype for the village ethics of hierarchy was the moral debt to his or her parents that a child acquired at birth (Malarney 2002: 112). This debt motivated some of the most enduring features of the villagers' ritual and ethical life. Funerals were of particular importance because they were the most public display of filial piety in village life. Reformers aimed to eliminate "superstitious" elements (Malarney 2002: 109–12) associated with the false ideas of causality that undermined the sense of ethical responsibility and historical efficacy a modern people should be acquiring. Funerals were also stripped of "wasteful" elements. There was a strong moral tone to this economic portrayal of ritual expenditure. The new socialist person should be a rational actor, understanding the utility of resources and using them for the common good. When a Ministry of Culture publication in 1975 exhorted, "In life my child gave me nothing to eat, in death the rice and meat become an oration for the flies" (quoted in Malarney 2002: 111), it was combining a materialist attack on the unreality of spirits with the rationalistic calculations of economic reason, reconstructing the model of virtue on utilitarian grounds.

The task of expanding the moral circle drew on the other distinctive features of communist self-understanding noted above. For example, villagers had to absorb the conceptual categories that would enable them to develop a sense of moral identification with large-scale social orders, such as nation, class, and oppressed people, regardless of their own more immediate and concrete experiences. This sense of identification would be strengthened by the expanded sense of agency that would follow the elimination of "superstition" and "fatalism." As Ho Chi Minh put it, "The people who are masters of their country must tend to the affairs of the nation as they do the affairs of their own home" (quoted in Malarney 2002: 55). The expansion of the moral circle and the fostering of an expanded sense of agency were bound up together. In practice, the attack on the family as the center of ethical thought took the form of abolishing or reconfiguring the rites of ancestor worship that enacted those obligations. Among the practices on the party's agenda in the village were hard work, maintaining good health, participating in communal practices, and incorruptibility, along with the reduction of expenditures on ritual, feasting, and gambling (Malarney 2002: 61–62). At the same time that the family was under attack, the idioms of kinship were taken up and expanded to facilitate emotional identification with those larger, more abstract groups. This is why "uncle" Ho Chi Minh encouraged the idea that he remained a bachelor because his family was the nation (Malarney 2002: 57; Marr 1984: 99).

REFORMING SOCIAL INTERACTION

Marxist-Leninist theories of class and history, and the categories of party, class, and nation, might provide the God's-eye perspective that would induce ethical reflexivity, but how could such big abstractions be made practicable in the concrete terms of everyday life? The more obviously economic and political aspects of the communist revolutions are well known: expropriation of large owners, land redistribution, forced collectivization, ideological indoctrination, and so forth. But the project of producing the socialist person dwelt just as much on the nature of ordinary interactions—the domain in which recognition and respect are most frequently accorded or denied. It was understood that material equality without the elimination of hierarchies of social value would be unstable or worse. For even if material inequalities were done away with for the present, hierarchies of social value would continue to motivate people to reproduce them once more. The revolutionaries strove to reform the semiotics of face-to-face interaction, which they recognized as fundamental to the old social hierarchies and the sense of an ethical self these involved.

According to Malarney (2002: 130–31, 165), villagers felt that hierarchy could be countered by "sentimental" relationships that minimize inequality; to lack sentiment was to live without virtue. Sentimental relations were expressed in everyday interactions: "Tremendous importance is placed on everyday forms of greeting between co-residents. . . . Thinh Liet residents say that those who do not greet others in fact disdain (*khinh*) them, thus the oft-quoted adage, 'a word of greeting is greater than a tray at a feast'" (Malarney 2002: 129). Recognizing the ethical power of ordinary interactions, some elite activists adopted new personal styles early in the century. One important anticolonialist, Phan Chu Trinh, denying deference to colonial authority was known for "his slashing manner of speaking, his refusal . . . to talk to the French prison director unless he was invited to sit down" (Marr 1971: 276). These were not merely superficial matters but were taken by observers to embody an ethical stance. When Ho Chi Minh stood before the crowds to declare the founding of the Democratic Republic on September 2, 1945, his comrade Vo Nguyen Giap recalled,

> the Old Man moved at a brisk pace. This surprised quite a few people then because they did not see in the President the slow and elegant gait of a "nobleman." He spoke with a faint accent of someone from the countryside of Nghe An province. Uncle appeared in this manner

> in front of one million people that day. His voice was calm, warm, precise and clear. It was not the eloquent voice people were used to hearing on solemn ceremonies.... It was full of life, and each word, each sentence pierced everyone's heart. Halfway through reading the Declaration of Independence, Uncle stopped and suddenly asked, "Can you hear what I'm saying?" One million people answered in unison, their voices resounding like thunder: "Y . . . es!!" From that moment, Uncle and the sea of people became one. (quoted in Marr 1984: 133)

We might think here of the aristocratic condescension of the Duke of Devonshire, described in chapter 5; but in this case, the egalitarian demeanor was quite programmatic and set some of the basic terms for revolutionary ethics by which interactions were to be judged throughout society. In his bodily and verbal demeanor, Ho served as a concrete moral exemplar of what might otherwise be highly abstract values.

Vietnamese language pragmatics had formalized a wide range of devices for establishing status and denoting moral relations. As in the European languages, the ability to make choices among pronouns (as in the French alternatives *tu* and *vous*) was a focus of status negotiation. But Vietnamese pronouns, working in coordination with a host of other linguistic devices, could also imply moral judgments. For example,

> if a speaker wanted to castigate someone, he might use *may* for "you" or *no*, *han*, or *'thang* for "he." . . . [W]riters could indicate their own attitude toward a particular character by their choice of pronoun. Historians, who had long divided historical personages into "legitimate" (*chinh*) and "spurious" (*nguy*), naturally selected pronouns that heightened the moral color of their narrative. (Marr 1984: 175)

Elite reformers had targeted the linguistics of hierarchy well before the communist victory. As noted above, the French addressed their employees with humiliating expressions. Moreover, they routinely ignored an important Vietnamese taboo on using the name, or even homonyms of the name, of someone senior to the speaker (Marr 1984: 181). By the mid-twentieth century many activists were practicing a self-conscious verbal egalitarianism. According to Marr

(1984: 173–74), the leader of a youth group would ask to be called *anh* (elder brother), and even Catholic priests might be called *cha* (father) rather than *co* (paternal great-grandfather). The honorific *bam* was dropped when addressing the powerful. Under communist rule, "persons of status demanded less language deference, and, more importantly, subordinates were far less willing to accord it to them" (Marr 1984: 173–74).

Saturating everyday interactions, the influence of language pragmatics was an inescapable feature of life across society. Like some of the apparently trivial matters brought up in the feminist consciousness-raising groups discussed in chapter 5, this inescapability was precisely why language reform was so important. Thus the assault on hierarchies in the village began not just with the redistribution of resources but with enormous attention to forms such as kowtowing and distinctions of dress (Malarney 2002: 34ff.). But language was a special concern. Honorific titles were prohibited. Self-deprecating modes of address were replaced by kinship terms or "comrade" (*dong chi*). In the early years after the 1945 revolution, hierarchy was reversed, and former landlords were supposed to be addressed by the degrading pronoun *may*; they in turn were supposed to address people with honorifics.

As the anthropologist Hy Van Luong shows, in both practice and theory, Vietnamese language pragmatics reflect the dynamics of deference and demeanor we saw in chapter 2. The use of person-referring terms is expected to shape the speakers' definition of the situation and therefore their patterns of behavior. These terms include both nouns and classifiers, words that modify nouns, for example, to exalt them; for example, to the already elevating term for "grandfather" one might add the further honorific for "holy" (Luong 1988: 241–42). Because these are words, or optional morphemes, and therefore work on a level of language where speakers are able to make choices (unlike more subtle grammatical distinctions), they are relatively salient to the participants' awareness. Therefore they are quite effective means of conveying people's evaluations of one another and of themselves. At the same time, because they are governed by habits of etiquette, they can seem to flow along naturally in the conversational background, in comparison to bold statements of praise or blame, full propositions that are more easily challenged as being false or insulting.

The result is a system of marking that, in Luong's (1988: 242) analysis, falls along two axes. One is a gradient of greater to lesser degrees of deference. The other implies the speaker's greater or lesser solidarity with the persons being spoken of. As Brown and Gilman (1960) show, in their classic study of deferential pronoun use in Europe, these two axes introduce an inherent ambiguity, since you might want to express solidarity or intimacy with someone you respect, such as your grandfather, or, conversely, distance from someone who is not owed respect, such as a worker. So in Vietnamese, a kinship term such as *anh* (elder brother) might also be used for an unrelated person. In such a case, it might either express the speaker's relative lack of respect for another person or emphasize the feelings of closeness between them. This ambiguity could open the door to reframing the social meanings of these signs. The word *thang* (male nobody), which might be used simply to avoid giving deference to the person being spoken of, was taken up by communist writers to express strong hostility.

Such apparently subtle verbal shifts have power, in part, because of the way they saturate conversation and give hierarchy its taken-for-granted quality. Both colonial and communist partisans were alert to their effects. Luong's (1988: 246–48) study of colonial-era writings found that whereas writers in Vietnamese who sided with the French tended to stress hierarchy in their linguistic choices, communists tended to stress the axis of solidarity. The colonial press used the terms for elder siblings for deference, and the communists, for solidarity; the latter tended to avoid the more hierarchical terms such as those for "grandparent." As the communists rose to power, they found in the existing verbal strategies for marking informal and egalitarian solidarity among peasants, such as sibling terms for non-kin, means to promulgate the aspirational habits of revolutionary ethics (Luong 1988: 249). By reconfiguring the boundaries of respectful behavior and its appropriate objects, language reform was also a key component in the effort to expand the range of moral concern beyond immediate kin and fellow villagers.

But, anchored in the resilience of verbal forms, the habits of interaction exerted a powerful drag on reformist efforts. Thus, Luong notes,

> even in moments of dominance [after the revolution], the linguistic usages of the Marxist movement were partly shaped by those of the former establishment: in the official Vietminh newspaper *Cuu Quoc*, Ho was referred to, not as *anh* (elder brother), but as *cu* (great-grandfather) and *chu tich* (Chairman).... In more informal contexts, the implication of solidary brotherhood was preserved between Ho and his followers, but through the use of the relatively more formal *bac* in reference to Ho (senior uncle to speaker's children) and Ho's use of *chu* and *co* to address his younger comrades (junior uncle and junior aunt hypothetically to Ho's [nonexistent] children).... It was a brotherhood marked by the distinction between senior and junior siblings, thus implying a hierarchical relation. (1988: 248)

This was not only the result of the conservative nature of habit, however. The problem also arose from a contradiction between two ethical imperatives, to express equality and solidarity, on the one hand, and to confer honor and deference, on the other. Thus, in everyday speech in the northern village Luong studied, by the 1980s, *comrade*, once a marker of revolutionary solidarity and equality, had come to imply formality and respect. The importance of these alterations in the minutiae of everyday speech norms is apparent in the censorship regimes, for both the French and the subsequent Marxist governments policed how people wrote and spoke in public (Luong 1988: 249). The inescapable irony, of course, is that the project of reconstructing social relations on antihierarchical grounds was directed by leaders whose authority and sense of ethical responsibility for society as such was partly based on Confucian moral hierarchies.[2]

THE VARIOUS FATES OF ETHICAL REVOLUTION

There can be no simple story to tell about the ethical revolution in Vietnam, even if we focus just on the regions where communism had its strongest roots. Before independence, a host of contending ethical ideas, values, and practices were distributed unevenly between small urban elites and a large rural peasantry. A heterogeneous range of

[2] I am grateful to Victor Lieberman for emphasizing this point to me.

factors, including the goodwill that communists accrued from their pivotal role in wresting independence from France, the strategic talents and charismatic leadership of Ho Chi Minh, land reforms and other economic benefits, and ideological work by party cadres at the local level, helped propel a localized version of revolutionary ethics to the fore against competitors. But from the declaration of independence in 1945 to the capitulation of Saigon in 1975 and Sino–Khmer war through the 1980s, the demands of warfare and the ever-present influence of the Soviet Union and China, along with the persistent economic trouble these involved and the challenge of assimilating former enemies in the south, would always constrain the capacities of Vietnamese, elite or otherwise, to be the full agents of their own social self-transformation (Duiker 1995; Taylor 2013). As an empirical phenomenon, ethical history can never be told independent of pragmatic possibilities and constraints of disparate origins, whose conjunction will be at least partly contingent.

In Thinh Liet, some people adopted the atheist position enthusiastically, and others continued to insist on the reality of the ancestors and their hand in the fortunes of the living; most, it seems, found themselves somewhere in the middle. In the 1990s, Malarney (2002: 102, 212–14) learned that younger males with party backgrounds were most likely to advocate the reform of ritual and older women were the main supporters both of the ancestor rituals of filial piety and of Buddhism. But not all reforms were equally provocative: the elimination of arranged marriage and of symbolic abasement in funerals was widely accepted. What provoked controversy were weddings, funerals, and death anniversaries, those rites most concerned with moral obligations to ancestors and ethical bonds among the living (Malarney 2002: 212–13). These rites remained flash points for ethical controversy. As the most public performances most people would carry out, the rites required stressful decisions about which revolutionary dictates to follow or ignore, forcing villagers to answer the questions "'Who am I?' A party member? A committed revolutionary? Or a loving mother? A devoted father? A filial son or daughter?" (Malarney 2002: 213). These were moments, somewhat like those imagined in the "trolley experiments" discussed in the introduction, at which people were forced to make clear, visible choices between competing definitions of the situation. It was precisely because of their salience, their capacity to define the situation and its actors, that these rites objectified people's ethical identities for others. At this point of giving an account of oneself in public, people's relations and identities were most subject to ethical evaluation.

Malarney reports that the result was a moral world divided between revolutionary and nonrevolutionary moral visions:

> The former envisioned a community that existed within a revolutionary nation-state. It was built around such principles as democracy, gender equality, "collective mastery," the equitable distribution of resources, rationality, science, party leadership, and commitment to revolutionary agendas. . . . The latter vision of the good society . . . centered on the "spirit of the village" (*tinh lang*) and its associated exhortations to "love with morals" (*song dao duc*). This vision valued honesty, respect, age-based hierarchy, mutual assistance, kinship ties, relations between co-residents, and affective "sentimental" relationships. (2002: 211)

But villagers were not just negotiating their way between a traditional vision and a revolutionary one. For one thing, the alternative to revolutionary ethics had long ceased to be merely a return to tradition. Moreover, the history the villagers had lived through was fraught with ethical troubles that forced them to reflect and innovate; top-down indoctrination by party cadres was hardly the only source of ethical reflection. As the anthropologist Heonik Kwon (2006) argues, the traumatic violence that Vietnamese villagers encountered forced them to grapple with a proliferation of novel, or formerly rare, kinds of morally troubling types of deaths—those of younger generations preceding their elders, of strangers and foreigners dying on village lands, and of kin dying in unknown locations—inducing them to invent new mortuary and memorial practices.

The death and disruption imposed by the massive violence of warfare undermined some of the practical means available for living a moral life. According to village traditions, one should ideally die at home and be buried amid kin and neighbors, remembered in altars, honored in periodic offerings. Given the conditions of warfare, which left vast numbers of the dead commingled with the remains of non-kin (including the bodies of American soldiers), left outside the village in distant battlefields, or lost altogether, the dead could become dangerous and disruptive hungry ghosts (Kwon 2006). Additionally, warfare, in which the young are likely to predecease their elders, who must then take charge of their obsequies, also reversed the proper order of obligation and responsibility.

Both filial piety and revolutionary ethics postulated a moral hierarchy of death. The former contrasted the good death (of an elder, at home, buried properly, ritually attended to thereafter) to bad death (sudden, violent, outside the village, ritually unresolved). The modern state added the category of heroic fallen soldiers (*liet si*) (Kwon 2006: 103). But none of these could deal with situations such as this, described by a survivor of a massacre in Ha Gia:

> When the floodwater subsided, we saw many skulls and bones. I did not know whether the skull belonged to my mother or my neighbor's sister-in-law. That's not a proper life; that's not a life at all. The authorities were not interested in correcting the situation. And I must be true to myself, too. We were busy with our daily life and trying to survive the hardship after the war. I tried to ignore the situation and tried to look the other way. Once in a while, particularly after the first lunar month [the traditional time of ancestor rites], the shame came back and induced unbearable pain in me. (quoted in Kwon 2006: 121)

According to Kwon, the villagers lacked a clear means of conceptualizing a moral resolution for bad death at such a massive scale, for "the victims were 'simple villagers,' and they were not. They were 'heroic defenders of the native land,' and they were not" (2006: 59).

Another side of the disruption of moral categories was an expansion of the circle of moral concern. As in the prewar era, villagers expected those who had suffered a bad death that was left ritually unresolved to become wandering ghosts. Decades after the gruesome massacre in My Lai, surviving residents were putting out regular offerings for the wandering ghosts of three American GIs, as well as that of a local mother grieving for her dead children. In another village, offerings were left for ghosts of anonymous fighters, one on each side of the war, a Vietminh and a Korean Marine (Kwon 2006: 101). We might surmise that actions such as these drew on deep sources in moral psychology, such as a universal capacity and propensity for empathy. But people are selective in the targets of their actions; their innate capacity for empathy is channeled by in-group loyalties, fear, and traditionalist habits. None of these factors alone is sufficient to explain the villagers' responses to their violent past, since the situation seems to summon forth all of them.[3] But when villagers improvised new responses to unfamiliar situations, they drew on the range of moral emotions, ethical vocabulary, and ritual work activities at hand.

HISTORY'S AFFORDANCES

As noted above, Ho Chi Minh said that ethics must be one foundation for socialist revolution. We also know that he was a canny strategist and quite attuned to his varied audiences, from North Vietnamese villagers and South Vietnamese urban capitalists to Chinese, American, and Soviet policy makers. So how should we understand his assertion? According to Marr, when Ho came of age, in the early decades of the twentieth century, young Vietnamese had been "bombarded with moral prescriptions both in and out of school" and Ho's

> ability to weld together revolutionary aspirations and selected traditional morality, often employing language that only a few years before had been the forte either of the Neo-Confucian literati or of the French-educated schoolteachers, surely represents one of the pillars upon which Ho Chi Minh's political accomplishments were built. (1984: 57)

Does this mean that we should take him to be a manipulative cynic? I think not. Marr is issuing a warning against literal-mindedness, not credulousness. Elsewhere he adds:

[3]Situations like these confound the ethical distinctions between face-to-face intimacy and abstractions. As ethnographer Erik Mueggler remarks about efforts by a minority group to cope with the aftermath of the Chinese Cultural Revolution, ghost stories "confronted the tortured question of how to apportion blame for starvation and violence between the distant, imagined center and one's own kin and neighbors," by treating the state as "both an indeterminate imagined entity that possessed actors . . . motivating them from afar, *and* a collection of determinate agents who bore moral responsibility for their actions" (2001: 269).

> The mere fact that Ho Chi Minh employed such terms as "loyalty," "humanness" and "virtue" did not make him a Confucian any more than Winston Churchill's or Franklin Roosevelt's use of such terms as "reason," "freedom," and "democracy" made them simple creatures of the Enlightenment. Each became a national leader in part because he knew how to take old symbols and use them creatively according to the political needs of the moment. (1984: 134)

The point is not that Churchill and Roosevelt were merely cynical and manipulative (politically canny though they surely were). Rather, it seems that Ho was just as earnest (and canny) as they were. But, of course, when terms are used in new contexts with such authority, their meanings are prone to bend and shift. The appearance of stability over time that an ethical vocabulary presents may be deceptive. Recall the feminist's "anger" or the Duke of Devonshire's "condescension" discussed in chapter 5. These words crystallize certain dimensions of ethical life such as emotional experiences or styles of interaction, in light of the specific sets of worries, interests, and projects predominant at a given historical moment. Times change, and so do people's major preoccupations. But the language changes at a different pace; I can still speak of "condescension" today, but certainly not in the eighteenth-century sense of the word. An existing ethical vocabulary may itself offer affordances—even invite new discoveries—to be taken up in new projects, as people respond to new concerns and possibilities in their changing circumstances.

Now all of this could make the business of taking up an existing ethical vocabulary seem quite instrumental. But to say that a leader such as Ho took old symbols and used them according to political needs, even if correct, can only be part of the story. For those needs, and the leader's insistence on them, were not simply given by the objective facts of the matter. We must also ask how those needs came to be what they were. Unless we assume that they can be fully accounted for with reference to some wholly self-explanatory objective terms (and what would they be? the drives of selfish genes? the social benefits of altruism? the forces of modern progress? the inner logic of class struggle? the magic of national destiny?), there is no Archimedean point where Ho, or Churchill, or Roosevelt can stand and view the scene with rational objectivity in order to freely take up and wield the tools he finds there. To be sure, I have argued that the viewpoints offered by such things as interaction with foreigners and education in moral and political theories can offer a secular equivalent of the God's-eye perspective. But even these remain embedded within particular historical contexts and are tied to certain values: they do not suffice to emancipate anyone from the constraints of psychological and social existence altogether. Rather than seizing on ethical vocabulary as a tool in order to accomplish some preformulated goals utterly apart from that ethics, Ho may well have been encountering potentialities latent in the ideas and institutions he was appropriating. If in the process he was altering those ideas and institutions, he might have been doing so at their invitation, discovering in them new possibilities.

Piety movements and socialist ethics find that everyday life with other people can be a problem. Whether focusing on individual salvation, like pious Urapmin and Egyptians, or identifying with large abstractions, like communists, people may well find that the ethics embedded in social interactions continues to press them. The pious cannot disentangle themselves entirely from the moral pull of kinship, neighborhood, and economic activities. Communists likewise may still have ancestors and ghosts to contend with, as well as the dignity and respect that the mere presence of their fellow villagers summons forth. Faced with Godwin's proposition, noted in the introduction, that one should transcend one's own place in the world in the name of a universal principle, saving the virtuous bishop rather than one's own mother, who is a mere chambermaid, one might balk. In the historical long run, it seems that these kinds of small-scale, intimate relations keep reasserting themselves. They may restore old particularisms or foster new bigotries, but they can also diffuse hatred. For instance, the routine expectations neighbors have of one another can sometimes undermine the demands of racial, religious, or political purism (Das 2007). In this way the moral circle may expand, but never fully escape the pull of those immediate others who embody the concreteness of ethical life. Even if those others are physically distant and electronically mediated, their demand remains local.

Having said that, a theme that does run through both the piety movements discussed in the previous chapter and the revolutionary one here is the way in which people in some historical circumstances came to be endowed with a compelling sense of ethical agency over themselves and their historical circumstances. That sense of agency was due, in part, to the historical objects available to them. It was these historical objects that helped grant them the God's-eye view. In the case of piety movements, of course, that perspective is identified with God as such, with the help of rituals, texts, preachers, and so forth. For Vietnamese communists, the God's-eye view was afforded by the texts of Confucian, Buddhist, and French moral pedagogy; the Marxist-Leninist theory of history and the social categories it presupposed; the institutions of modern nation-states and political parties; and the ongoing need to take the perspective of hostile and, occasionally, friendly outsider powers on their own local circumstances. These perspectives fostered and facilitated a sense of agency whose realization, of course, was affected by heterogeneous and contingent factors. If that history can never be told only in terms of the unfolding of ethical motives and goals, neither can we make sense of it without them.

Conclusion

AFFORDANCES, AWARENESS, AGENCY

Finally!

Motivating this book is the effort to gain insight into ethical life by moving among the perspectives offered by natural and social histories. I have argued that each provides a partial context for the other, but neither fully determines the ethical life that results for individuals or their communities. Each of the various research traditions we have surveyed grasps certain crucial dimensions of ethical life. To say this presumes that these different fields are, in some nontrivial respect, talking about the same world. This is not a view that all their practitioners necessarily accept. The more historically oriented fields, such as anthropology and social history, often take those dimensions explored by psychological and linguistic research as background so deep that it can be safely ignored or, more negatively, as a source of spurious reductionism. Naturalistic approaches sometimes treat the more historical domains as merely exemplifying something they already know that contributes no new information or, worse, as misleading epiphenomena to be seen through. The result can be either mutual indifference or antagonism. That antagonism is sometimes justified, when proponents of one field proclaim that they have the unifying explanation that will dissolve all other approaches within a single solution. Such totalizing ambitions have not fared well: even particle physics does not explain why square pegs don't fit in round holes.[1] Mutual indifference can blind us to the insights we might gain not from the stratospheric viewpoint of unified theory but closer in, by crossing the borders of adjacent fields. These border crossings should move between natural and social histories in both directions.

[1] At least if those pegs and holes are not the size of protons (David Gerdes, personal communication, November 13, 2014).

As I argued in chapter 4, if the sources of the emotions lie in universals of human neurophysiology, their distinctively ethical character is only brought into focus through the dynamics of social interactions. These interactions draw on moral categories such as "condescension" or "male chauvinism" that emerge out of specific histories, such as those sketched in chapter 5. Moving in the other direction, chapter 3 argued that Melanesians' claims about being unable to read one another's minds can only be fully grasped in all their ethnographic specificity if we realize that they are grappling with challenges posed by capacities such as empathy, intention-seeking, and reciprocity of perspectives shared by humans in general, as discussed in chapter 1. Against an entirely deterministic explanation of the relations between the different dimensions of ethical life, I have suggested that we include a role for affordances. The idea of affordance, as argued in the introduction, acknowledges both naturalistic realism (there is a neurophysiological "there" there) and historical creativity (Athenian or Confucian virtue and the abolition of slavery could each have turned out differently). It suggests how they might interact.

Here I will briefly summarize the argument of the book and then turn to some puzzles posed by some of the strongest present-day expressions of the idea of the expanding moral circle: the globalization of human rights and humanitarian movements. In the discussion of the natural histories of ethics in part 1, we paid particular attention to the evidence from child development. There we found a range of capacities and propensities that seem to appear in one way or another in children across all societies and are important elements of mature adulthood. Among them are gut feelings of disgust and attraction, empathic impulses, propensities to evaluate other persons and discriminate among them, spontaneous sharing and cooperation, notions of fairness, intention-seeking, reciprocity of perspectives, self-distancing, and a tendency to seek out, conform to, and enforce norms. I proposed that these do not in themselves fully qualify for the label "ethical," nor do they necessarily give rise to ethics. Indeed, there is no consensus among evolutionary theorists that these capacities and propensities are designed to fulfill any particular social functions. But they are routinely recruited, and sometimes denied, in the service of ethical life.

In part 2, we then looked at the social dynamics in which people encounter one another in (mostly) face-to-face interaction. The focus here is on the reciprocity between first- and second-person perspectives, something codified in the pronoun system every language must possess. As we saw in chapter 1, this capacity for shifting points of view between self and other is fundamental to human cognitive development (Tomasello 1999)—and the neurological collapsing of the distinction altogether may be a spur to altruistic action (Pfaff 2007). In part 2 we found certain patterns that seem to be fundamental features of social existence anywhere. These include basic conversational techniques such as turn-taking and repair, as well as the ways these become socially consequential, for example, in distributed agency, imputations and denials of responsibility, face-work, stance-taking, or the typifying of people's actions and character, for instance, through gossip. Again, these are not always and everywhere "ethical." Some seem to function primarily as technical coordination devices. But they too offer affordances on which ethics can draw. For example, as noted in chapter 3, Darwall's (2006) analysis of the concept of "dignity" requires address to a second person, the demand that "you" respect "my" dignity. The ethical potential of techniques of social interaction may be brought into focus and elaborated in culturally specific ideas expressed in words such as the English *condescension*, the Sumbanese *dewa*, El Barrio's *respect*, the Chewong *punén*, and the Inuit *ihumaquqtuuq*; in thematic patterns such as Korowai worries about mind reading; or in Don Gabriel's deployment of distinct speaking styles or voices. Such ideas are not direct products of human psychology or social interaction; rather, they result from particular social histories that have responded to affordances of psychology and interaction. It is in the second-person perspective that people are compelled to account for themselves to one another—for example, with excuses, accusations, justifications, praise, or blame. Sometimes these acts are quite subtle, but they are prone to entering into processes of objectification, ways of making things explicit and thus of making ethics a matter of heightened awareness. In the process of doing so, they draw on the third-person perspective, the point of view that speaks in the generic terms of a given ethical vocabulary. For it is the character of words such as *dignity*, *respect*, *dewa*, *punén*, and *ihumaquqtuuq* to convey their meanings into an indefinite number of possible contexts, regardless of who "I" or "you" may be. In this function, they facilitate such basic psychological propensities as third-party norm enforcement (Kalish 2012; Kalish and Lawson 2008; Moll and Tomasello 2007). Cognitive and semiotic resources such as these seem to be preconditions for the general, potentially context-free, norms so characteristic of morality systems.

I have argued that social interaction is the natural home for ethical reasoning, explanations, justifications, and all the other dynamics that prompt reflexivity. By examining these dynamics, we can begin to respond to the challenge that the psychologists have posed to ethical thought. The challenge, put briefly, is this: If people do not have the consistency of character they think they have, and if their acts are not due to the reasons they give for them, but character and action are shaped by processes that lie beyond their awareness, then what role does awareness play in ethical life? Part of the answer lies in social interaction. Even if an individual's actions are due to deep physical and psychological causes below the level of his or her consciousness, those causes only explain what that particular individual has done in a certain context. They do not entirely explain norms, patterns of reaction, judgments that circulate across a community, or act descriptions that endure from one generation to the next. It is in circumstances such as second-person address that awareness comes into play. When people account for themselves, or when they must cooperate, or when they seek to influence actions in the future, they must in some sense understand one another. Because their efforts at understanding depend on sense perception, they are semiotically mediated. Anything perceptible, including noises, gestures, images, or artifacts, will do, but given its special cognitive and pragmatic effectiveness, language plays a central role. And the explicit languages of ethics bring their histories into the scene of interaction.

Talk affects not only how one person interprets another but also his or her self-perception as well. As noted in chapter 3, there is evidence that an individual's neurophysiological state may not determine the exact emotion he or she experiences—for instance, the same stimulus might prompt anger or exhilaration. In the original experiments with epinephrine injections (Schachter and Singer

1962), the difference resulted from how the subject's companion behaved: being with a clown provoked emotions of euphoria; the company of an angry person induced anger. But the historical transformations described in chapter 5 suggest that the locally available categories and descriptions—such as "condescension," "humanity," or "sexism"—can play a similar role in shaping emotional experiences (cf. Cohen et al. 1996). Moreover, by giving feelings an object (such as anger about injustice), a particular vocabulary may play a critical role in transforming a physical state into a distinctively moral emotion. This suggests that there cannot in principle be a generic or universal moral emotion such as anger, since it will necessarily be shaped by historically specific prompts, targets, and descriptions.

Explicit ways of talking have been central to much of the ethnography of ethics and morality. Often focused on morality systems, cultural accounts have shown the range of norms, customs, laws, theological and philosophical principles, disciplinary practices, and verbal justifications these can take. If cultural accounts sometimes risk taking these explicit systems too much at face value, explicit objectifications nonetheless do play a crucial mediating role between natural and social histories of ethics.

When feminists engaged in consciousness-raising (chapter 5), they were taking advantage of the dynamics of conversational interaction, such as emergent patterning, in order to bring into view new objects of knowledge. Giving names to experience, for example, "woman's anger" or "sexism" or "date rape," had two immediate aims. One was to facilitate individual self-awareness; the other was to create a public world in which an indefinite number of other individuals would be aware of the same things. That is, different people could identify qualitatively *different* incidents as instances of the *same* ethical category. In the process, consciousness-raising groups were engaged in what the developmental psychologists call representational redescription, giving feelings a distinct target, for instance, something they could be angry *about*. As we have seen, this quality of justifiable "aboutness," according to Gibbard (1990), is part of what defines an emotion as being moral.

The feminists' political goals depended, in part, on the ethical work of redescribing everyday interactions so that other people could see the harm they entailed and the ways of flourishing they forestalled. As these redescriptions took off, some became historical objects. They became widely recognizable (I can count on you to understand me when I refer to "sexism" whether you agree or not), and people could actively respond to them (you may devote yourself to ending "sexism"—or deny that there is any such thing). Notice that although feminist consciousness-raising took advantage of the second-person perspective of face-to-face interaction, the outcome was categories whose sociological character entails the third-person perspective. Words such as *sexism* invite individuals to see their first-person, subjective experience from the distant perspective of a sociological category. This, in fact, is part of why the sessions were supposed to be therapeutic, as well as political—the group participant realized that what had seemed to be idiosyncratic subjective experience was not peculiarly her own, in isolation, but one instance of a general type. Rather than being alienating, this typification was meant to absolve her from taking her unhappiness to be either individual psychopathology or a failure of personal responsibility. In effect, this is a parallel to the exculpatory logic we saw in chapter 5 when Charlson was found not guilty of murder due to his brain tumor.

With the third-person perspective and efforts to change existing norms, we come to the topic of part 3 of this book, social histories. Here our attention is directed at purposeful endeavors to live up to or transform an ethical world. Explicit objectifications play a critical facilitating role in these endeavors. The various aspects of ethical life surveyed in parts 1 and 2 of this book—everything from reciprocity of perspectives, intention-seeking, and norm-seeking to concepts of dignity, verbal stylistics, and consciousness-raising sessions—are quite heterogeneous. They are not inherently ethical in themselves, nor taken together do they necessarily form a cohesive set. Something more is needed to induce people to recognize discrete actions and impulses as being ethical matters at all. That recognition is not usually something that any single person does on his or her own. Like feminists engaged in consciousness-raising, when Korowai worry about other people's intentions (chapter 3) or Don Gabriel struggles with the inner clash of voices (chapter 4), they are drawn into the processes that may induce some form of ethical self-consciousness.

That self-consciousness may be evanescent, or it may become socially successful, resulting in ways of describing and evaluating people and their actions that *other* members of the community also recognize as distinctively ethical.

It is important to bear in mind that ethical life is not defined by self-awareness in any straightforward way. The dynamics of social interaction sketched in chapter 2, for example, or the ethical stances identified with Don Gabriel's "voices" in chapter 4, seem rarely to intrude on anyone's consciousness. But that consciousness does play an important role in the social history of ethics. For the process of consolidation and objectification is what gives order to the heterogeneous materials of ethical life, allowing people to agree with one another that there are ethical stakes in, say, reading intentions (for Korowai) or profit-seeking (for Don Gabriel). At this point, categories and descriptions can be recruited into social projects or by individuals trying to cultivate capacities such as self-distancing and norm-seeking, whose expressions include the Christian and Muslim piety movements discussed in chapter 6 and the Vietnamese communist revolution in chapter 7. But if those purposeful efforts are to be more than private thought experiments, they will inevitably become entangled with economic, political, cultural, and social dynamics that confound any attempt to reduce ethical history to a single story line. As we saw, Urapmin Baptists must still face their kinfolk and worry about their place in the larger world, followers of Muslim piety movements must wrestle with everything from authoritarian states to the ordinary demands of making a living, and Vietnamese communists had to defend themselves against violent foreign attacks, deal with the unwanted demands of political allies, and after victory, coexist with apolitical or anticommunist fellow citizens.

The social movements sketched out in part 3 have two general features in common. One is their link to some idea of progress; the second is their social breadth. Abolitionism, feminism, born-again piety, and communist revolutions are not only purposeful efforts by people to exert agency over their ethical worlds. They are also oriented by some sense that history has—or at least could have—a direction, that there is ethical progress. That progress is supposed to be based, in part, on people's growing self-awareness and the grasp of historical agency that awareness, facilitated by laws and institutions, should foster. The result is sometimes described as an ever-expanding moral circle that tends toward universality (for versions of this view, see Bloom 2004; Pinker 2011; Singer 1981). Abolitionism extended to enslaved Africans the ethical concern formerly confined to white Europeans, feminism (in the first instance) sought a similar extension from males to females, Christianity and Islam (in principle) expect all humans to be capable of salvation, and communism expected eventually to achieve a global outcome.

Two important contemporary international movements in the nonrevolutionary secular world presuppose some version of ethical progress in the form of universalization. The human rights movement aims to realize the assertion in the 1948 United Nations Universal Declaration of Human Rights that "all human beings are born free and equal in dignity and rights" (Hunt 2007: 17). The second movement is humanitarianism. Although often linked to the first, humanitarianism tends to focus on suffering and the prevention or amelioration of physical harm. Both movements, however, are predicated on ethical universality in principle and its global reach in practice. That is, since ethical values, the sentiments they should induce, and the obligations they impose pertain to all humans, so too should ethical agency be indifferent to any distinctions of culture, ethnicity, religion, gender, age, or political divisions. A brief glance at human rights and humanitarianism will also display some of the tensions inherent in the conjunction of natural and social histories in ethical life.

HUMAN RIGHTS

The historian Lynn Hunt (2007) provides a good example of the mutual contextualization of natural and social histories in her narrative of the origins of the idea of human rights in eighteenth-century Europe and the American colonies. Recall the sociologists, quoted in chapter 6, who write of human rights that "everyone now agrees that human beings have such rights. In contrast, no one can say from whence these human rights derive or on what they are founded" (Porpora et al. 2013: 4). This book should make clear why their diagnosis is off target: we should not expect to discover that the ethical life can only stand if it is provided with the firm foundations of a cohesive set of explicit justifications, religious or otherwise. Justifications start in medias res: they reflect back on an ethical life that is already in motion. If Hunt is right, this is true even of the sources of current doctrines of human rights. She argues that changes in a heterogeneous range of everyday social practices in that era helped give plausibility to new concepts of rights. Among these were changing sensibilities about the bounded body (not spitting or urinating in public, eating with utensils instead of hands) and the separateness of individuals from one another (sleeping alone or with a spouse, listening to plays and concerts in silence), as well as new doubts about the authority parents should hold over their children. In the same period, certain arts elaborated aspects of moral psychology that they had previously ignored: portrait painting directed viewers' attention to individual personalities, and novels encouraged their readers to focus on the individual's interiority and to empathize with people of different classes and genders, such as female servants.

None of these developments in itself is necessarily ethical in character, nor do any of them form a clear-cut set with a single defining essence. Anthony Appiah captures a facet of this in his reflections on the trolley experiment described in the introduction. He comments,

> In the real world, situations are not bundled together with options. In the real world, the act of framing—the act of describing a situation, and thus determining that there's a decision to be made—is itself a moral task. It's often *the* moral task. Learning how to recognize what is and isn't an option is part of our ethical development. (2008: 196)

Here he is referring to the artificiality of the trolley problem, which presents itself to the individual as a multiple-choice test: Push the fat man, or divert the train, or do nothing. But his remark contains a broader insight as well: it points to the business of defining something *as even being ethical at all*, and not, say, just a question of efficiency.

Consider the wide and quite heterogeneous range of practices we have looked at in this book, which include the child's cooperative impulse, conversational repair, gossip, didactic instruction, consciousness-raising sessions, religious disciplines, and political revolution. As I have been arguing, none of these is necessarily ethical—each might be described as the means to some immediate, practical, and value-neutral end. This is true of the eighteenth-century practices that Hunt identifies as the context within which human rights came to make sense, even to seem necessary. Eating with utensils could be explained as a matter of hygiene, limits to parental authority might simply have been a response to new psychological discoveries, and silent audiences were perhaps just adapting to changing aesthetic fashions. These developments do not of necessity have to be seen as ethically inflected. Further unifying processes had to occur if feelings about bodily decorum, privacy, autonomy, and sentimentality were to coalesce into a particular way of life that encouraged people to see them in distinctly ethical terms. To have the historical impact that Hunt describes, those developments cannot simply result from individuals engaging in inner reflection. That impact is the outcome of distinctive dynamics of reasoning, justification, new habits, institution building, and social movements. It is only in light of such dynamics that the new sensibilities could be consolidated in such a way that they come to be *recognizable* as ethical. Being recognizable means that they have a public life in which one individual can expect that others will also grasp their ethical character, even if they differ over what to make of it. People in the eighteenth century could certainly bitterly disagree about human equality (those storming the Bastille, for instance, versus those destined for the guillotine) but might still acknowledge that the argument was in part an ethical one. They would recognize that they were arguing about something more than practical utility or forms of governance (although of course each of these might also be involved).

Natural and social histories unfold in quite different scales of time, from the slow movement of an evolution that has no master plan to the sometimes precipitous doings of individuals with goals. Taken together, the assertions of natural and social histories can give rise to paradoxes. Hunt points to one: among the founding expressions of the idea of human rights, in her view, is the American Declaration of Independence, written in 1776, which says of them that "we hold these truths to be self-evident." Noting that the ideas expressed in the declaration arose in the Euro-American world of the eighteenth century, Hunt remarks, "If equality of rights is so self-evident, then why did this assertion have to [be] made and why was it only made in specific times and places?" (2007: 19). In part, this is a question about progress and universality. As noted in chapter 1, children seem to develop gut-level intuitions about fairness early in their development: So how can something that is purportedly universal also be subject to change? If it is both universal and self-evident, why had it not been noticed before? For one thing, the innate sense of fairness runs up against another feature of child development, the propensity to divide people into "us" and "them," with moral obligations applying only to the first (Hirschfeld and Gelman 1997; Rhodes and Chalik 2013). If fairness and discrimination are core components of the natural history of ethical life, their shifting application and distribution, and the way one (such as fairness) may be brought to bear on altering another (such as discrimination), is an outcome of social history.

Hunt's story, however, is also about the actual consequences of assertions, of making things explicit, of the dynamics that link objectifications—such as a declaration—and ethical awareness. If the psychologists we reviewed in chapter 1 are right, those consequences might not affect people's cognitions or emotions directly. They have given us strong evidence that the reasons people give for their decisions or their ways of life do not correspond with the causes and contexts that most directly affect them. Rather, making ethical ideas explicit brings about changes in the public context within which cognitions and emotions are developed and expressed. They alter the slope of the land in ways that channel the flow of ideas. In any given context, some concepts and arguments make more sense to people, and others, less so. The crucial point here is that these ideas enter into social interactions. If we only look at individual reflection, all sorts of possible ethical worlds are imaginable, and in theory any given individual may come up with new ones. But these ideas face the test of other people: to them, some of my ideas will seem self-evident, but others may remain utterly incomprehensible. If ethical revolutions sometimes occur, it is in part because what other people will find recognizable has changed.

Hunt moves from the American Declaration of Independence to the Declaration of the Rights of Man and Citizen in 1789, during the French Revolution. She says that making ideas about rights explicit had two results. First, it changed the parameters of what was thinkable (or perhaps I should say, what could be openly said and readily understood). Once values such as equality were made explicit, they became available for conceptual elaboration in ways that no one could predict and drew on underlying, and until then tacit, conceptual categories: "The logic of the process determined that as soon as a highly conceivable group came up for discussion (propertied males, Protestants), those in the same kind of category but located lower on the conceivability scale (propertyless males, Jews) would inevitably appear on the agenda" (Hunt 2007: 150). The logic is not inherent in the underlying categories all by themselves; it depends on the way they came up for discussion. The logic of the process is both cognitive and social in nature. It results from the cognitive effects of making things explicit, subject to heightened awareness. They are thereby more readily subject to the social effects of entering into the purposeful activities of people who are talking with one another, because they are working with (or against) one another.

To repeat my point: the natural historians are right when they cast doubt on the reasons people give for their actions or their perceptions of character; these reasons may not be the same as the causes underlying actions or the appearance of character traits. But the social historians are right to insist that consciousness and purposes matter. (They themselves may, of course, be skeptical of people's self-justifications for different reasons, such as ideological biases.) To understand this, I have argued that we cannot leap straight from psychological or other capacities to social history. There is a crucial middle ground, the domain of social interaction, where people see themselves through one another's eyes and must account for themselves to one another. Sometimes they will discover something new about themselves in the process. I have summarized the distinct viewpoints as first-, second-, and third-person positions. Certain fundamental capacities and propensities, such as disgust and approval, are fairly well captured from the first-person perspective of subjective experience. Others, such as empathy, sharing, cooperation, intention-seeking, and reciprocity of perspectives, require what I have called a second-person perspective. They involve address to another person, and just as important, they (usually) take that other person as someone who addresses me. Interaction puts me in the positions of both first and second person. Fully developed, therefore, these perspectives depend on a degree of self-distancing: I must have some ability to see myself as you do. This is not just a cognitive skill, for instance, knowing what I look like from the outside. Early on in life, this second-person perspective gets built into the rapid, instinctual rhythms between child and caretaker, on which is built the language learner's facility for pronoun switching and conversational turn-taking. The indefinitely many different styles of social interaction among adults (which vary across cultures and, within them, across registers) develop and focus basic psychological and linguistic capacities and propensities. In other words, the way in which selves come to have ethical standing on an everyday, ongoing basis depends in part on the mutual regard, or disregard, that transpires in the second-person address between them. What virtue ethics calls the cultivation of the self depends on this mutual relationship between first and second persons. What the promptings of psychology suggest, and the explicit dictates of morality systems teach, can only be taken up to become a way of life by people who are enmeshed with one another on an ongoing basis.

Interaction itself chronically brings in the third-person perspective. Again, this perspective draws on capacities and propensities that appear very early in childhood, such as norm-seeking, conformity, and self-distancing. The second person is someone who addresses me and whom I address. This is the interactive space within which people find themselves giving accounts: justifying, excusing, accusing, explaining, denying, praising, blaming, and all the other activities in which ethical categories and stances are made explicit. These categories contribute to, and in turn draw upon, a public reservoir of concepts. The Polish peasant says of the Jewish girl, "She's not a dog, after all." The Kluane hunter recalls a grandmother's reminder to thank the rabbit for its gift. My neighbor Sally is struck by the force of an essay she read on a political blog. Or a neurologist persuades a jury that an apparently evil act—Charlson murdering his son—can be explained by the mechanical effects of a brain tumor, rather than the ethical choices of a free agent (see Laidlaw 2014: 193; cf. Neiman 2002). All of these acts draw on the third-person perspective, not the view from a person who addresses me but, rather, of the observer outside the action. They make use of ideas that, in principle, anyone who shares my vocabulary ought to grasp. This is the public world in which justifications and arguments make a difference.

Justifications and arguments make a difference because they are addressed to other people who can be expected to recognize their implications. One way people make history is by discovering in their arguments aspects that they had not noticed before, or considered relevant, and drawing from them new conclusions. As Hunt points out, once the French granted civil rights to Protestants, it became more self-evident why it would make sense to extend them to Jews as well. Or take the idea of fairness, which apparently emerges in early childhood and is one candidate for inclusion on Shweder, Mahapatra, and Miller's (1990) list of universal ethical values. According to the anthropologist Caroline Humphrey, eighteenth-century Mongols took for granted a social world that was crosscut by deeply felt principles of distinction and hierarchy. With differentiation by clan membership and ranked birth order, no one was equal to anyone else. People of various statuses differed not merely legally but in their very essence; they were considered to be fundamentally incommensurable. Yet Mongols were preoccupied by notions of fairness. What fairness usually meant, however, was treatment in accord to one's given status—to be fair meant granting people their dues in accord with their respective ranks and clans, which were quite distinct from one another. But, Humphrey tells us, this hierarchical concept of fairness could be applied more broadly. Thus in 1788, a minor official complained to his superiors who had enslaved some serfs and confiscated others' livestock, writing, "The suffering of the many is my suffering, and when it comes to miserable me, I can do no more than . . . send [this petition] upward" (quoted in Humphrey 2012: 315). Humphrey comments on this petition that "it goes beyond the fairness that would concede that others should have their due. Janggi Ashig puts himself in their position. He invokes not law (*hauli*) but 'norm' (*jirum*)" (2012: 315). In effect, Janggi Ashig takes up an existing language of fairness and norms and finds new possibilities in it. Humphrey does not tell us what led to Janggi Ashig's individual ethical insight, but the crucial thing is that he depends on this reworking of the available ethical vocabulary to be recognizable to the superiors he is addressing. To account for this requires both second- and third-person perspectives. He faces his superior as someone to be addressed in the second person and who in turn might address him. (Here it matters that Janggi Ashig's rank endowed him with a voice, something a serf would presumably lack.) But he counts on something else, that the shared vocabulary of fairness and norms will be recognizable to his addressee—not because it is he, Janggi Ashig, who deploys it but because these are categories that *anyone* who shares their public world should understand, even viewed from the third-person perspective of someone outside the interaction.

Janggi Ashig's innovative broadening of an existing notion of fairness might seem to be a comforting story of ethical progress, one step in an inevitably expanding moral circle. But that is not my point. For one thing, affordances determine no particular outcome. For another, as the previous three chapters show, social histories confound heterogeneous processes with results that do not display any obvious directionality. Consider again the concept of human rights. These usually include freedom of religion. Several scholars of law and politics have argued that freedom of religion leads to paradoxical results. The law professor Winnifred Sullivan (2005) shows how the legal enforcement of this principle requires that some institutions be able to define what religious liberty is and impose that definition on people. In doing so, she says, the law cannot help but favor some values and concepts over others. International law governing religious liberty has tended to assume a distinctly Protestant model of faith, for instance, as a matter of the inner belief of individuals rather than the external rituals of communities (Mahmood 2012). Treating the right to religious freedom as universal, the principle authorizes some institutions to intervene in the workings of others and supports some ways of being religious at the expense of others. Put another way, an expansion of rights in one domain may unavoidably require a diminishment of rights, such as self-determination, in another. This outcome is not necessarily due to malign intentions or hidden purposes. At the very least, it simply shows that once ethics has a public existence and an institutional life, it cannot be reduced to a single line of normative justifications or one set of psychological preconditions.

This example is subject to debate; I mention it here simply to make the following point. Those who propose that there is a history of ethical progress, of an expanding moral circle, cannot, I think, be describing the growing enlightenment of individuals taken in isolation. The development of institutions must play a crucial role in this history. As we saw in chapter 5, the British abolitionist movement depended on networks of churches to get under way in its early stages and on parliamentary law for its fulfillment. These became part of the background that endowed new norms with the taken-for-granted character they held for children who came of age a generation or so later. The ideals of equality and justice that inspired many of the Vietnamese revolutionaries discussed in chapter 7 were promulgated through international networks and enforced through state and military interventions. The globalization of human rights depends on institutions (the International Court of Justice, the United Nations, and the nation-states that sustain them and the nongovernmental organizations that work beside them) whose workings must respond not just to ethical reasoning but to political pressure, bureaucratic logic, funding constraints, and so forth. As a result, each of these cases, to varying degrees, contains its ironies, contradictions, disappointments, and even unanticipated successes. This is not just because there are constraints on progress but because ethical history moves along more than one dimension.

HUMANITARIANISM

Whether or not modern history does reveal an expansion of the moral circle, *the idea* that it does, or that it ought to, has had an impact. The concept of universal human rights (and its historical distinctiveness as being a mark of enlightened modernity) has given rise to nongovernmental organizations, the World Court, and other institutions. A related but distinct set of institutions has developed around the idea of humanitarianism (whose genealogical roots include the British abolitionist movement discussed in chapter 5). The anthropologist and medical doctor Didier Fassin, reflecting on his own long involvement with humanitarian agencies, including Médecins Sans Frontières, brings out some of the problems faced by the attempt to bring about a global ethical order. First he notes that the word *humanitarian* conflates two meanings. It refers both to the idea that there is a single shared human condition and to the feelings that draw humans toward one another (Fassin 2012: 2). Humanitarianism tends to treat the first as a condition for the second, indicating that the mere fact of being human is sufficient to warrant interventions in the lives of distant strangers. Although one might consider both of these to rest on the basis of empathy and other elements of moral psychology, Fassin looks elsewhere for their sources. In his view, contemporary humanitarianism is a secularization of the Christian notion of suffering as redemption, since "with the entry of suffering into politics, we might say that salvation emanates not through the passion one endures, but through the compassion one feels" (2012: 250). Again, this is not a view accepted by all observers (it might be rejected by Jewish, Muslim, and Buddhist relief groups, for instance), but it does point us toward the central paradox. Humanitarian groups are sometimes met with hostility and resentment on the part of those they are helping. One reason might be that their view of universal values is parochial. Fassin, however, points to another: they engage in altruism that offers little scope for reciprocity.

Seeing all humanity as one, humanitarianism presupposes a version of the God's-eye view, one in which there should be no distinction between suffering in one's own neighborhood and that in Darfur or Haiti. But that perspective is highly asymmetrical. It is, after all, the humanitarian who decides who merits intervention and whether or not to act. As Fassin points out, in order to remain recognizable, the recipients of humanitarian aid must meet terms set by the humanitarian. Referring to large-scale relief activities, Fassin says that humanitarianism reifies victimhood and pays more attention to the biological life of those who are suffering "than to their biographical life, the life through which they could, independently, give a meaning to their own existence" (2012: 254). They must be suffering, innocent victims, not pious Hindus, clever cattle raiders, resourceful mechanics, good daughters, or wounded insurgents—and they should certainly not turn out to be nothing more than poor people hoping to gain access to Western riches (Englund 2008).

The asymmetry between humanitarians and aid recipients becomes especially stark when we consider the very different treatment that Europeans accorded refugees in distant places and asylum seekers at their own borders at the beginning of the twenty-first century. According to Fassin, the refugees can be readily understood in terms of their suffering. A large undifferentiated mass, located at a distance, they fit relatively easily into the category that permits humanitarian relief: they can be grasped from the third-person perspective. Asylum seekers, on the other hand, are carefully sorted out as individuals, who are compelled to prove their worthiness of asylum on a case-by-case basis. In the position of persons addressing their potential rescuers in the second person, they are typically more subject to suspicion and rejection, at least under the legal systems in place when Fassin was writing.

Fassin's purpose is not to discredit humanitarian interventions. He acknowledges both their virtuous motives and their valuable consequences. But as with other social movements, their paradoxes and limits reveal something more general about the problems of ethical life. Referring to "the deployment of moral sentiments in contemporary politics," Fassin writes, "Through the moral sense it credits us with, it endows us with our own share of humanity. We become more fully human via the manner in which we treat our fellows" (2012: 1, 253). Here writ large we can see a version of the dynamic of self and other described earlier in this book, in which the way I treat you does not just express whom I take you to be but reflects on my own ethical character—a character that requires your recognition in order to come into its own. But notice the asymmetry: the humanitarian has to accord recognition on him- or herself, because, to the extent that the victims lack a voice, there is no one else fully able to recognize him or her. The moral self-awareness in this case depends on an imagined other. Compare this with some other forms of second-person address we have encountered in previous chapters. One is described by Elizabeth Anderson (2013) in her analysis of social movements mentioned in chapter 5. Anderson argues that social movements often bring about ethical change by forcing more powerful groups to take their interests seriously. It is those social movements, not the powerful, that take the initiative, and what they say is not predetermined (at least not entirely) by what those whom they address already think they know. In Darwall's (2006) version of second-person ethics, the value of dignity is sustained between people who can hold one another accountable. But as we have seen, in ordinary conversation the second person is not just a form of address and does not stand alone. It is predicated on the ability to alternate with the first person. First- and second-person linguistic forms are the semiotic armature for the reciprocity of perspectives so crucial to early child development. We might say that the ethical dilemma of humanitarianism is that, although it delineates the obligations of humanitarian agents, it falls short of realizing this reciprocity of perspectives, if in the end it is unable to grant the authority of address to the victims or offer instructions for how to be an ethical subject sustained within particular relationships to the humanitarians. Like many morality systems, it functions best at the plane of general principles, viewed from a third-person perspective.

FIRST-, SECOND-, AND THIRD-PERSON POSITIONS

The tension between the first/second- and third-person perspectives is exemplified by the challenge we encountered in the introduction, with William Godwin's question of whether one should save the bishop or the chambermaid from a burning house. In the same passage, considering that the chambermaid might be one's own mother or wife and yet one should still save the bishop, Godwin remarks, "What magic is there in the pronoun 'my,' to overturn the decisions of everlasting truth?" (1793: 83). What could be a more powerful expression of the third-person perspective than this, which completely eliminates the first-person perspective grounded in one's immediate psychological dispositions and social location and the second-person relationship to specific others in favor of a matter of rational principle? Notice that neither the bishop nor the chambermaid speaks up or has any ethical role to play, apart from being the object of my act of intervention. Asked whom to save, as suggested earlier, the utilitarian might favor one answer, the compelling obligation of the greater good, and the virtue ethicists, another, perhaps the value of relationships and their importance for constituting an ethical self over the long term.

But the empirical researchers in natural and social histories are not obliged to make this decision. Their findings need not aspire to either provide ethics with foundations or guarantee outcomes—other, perhaps, than to insist that we should expect ethics to crop up in all sorts of places and guises. Empirical researchers cannot simply exclude by fiat ethical worlds in which abortion is murder, members of different clans have distinct essential beings, or humans are subordinate to divinities. They cannot expect that the facts will ultimately assure us that all people will define the good in ways we recognize or that they will eventually converge on values that we can accept (see Doris and Plakias 2008). Does this, then, condemn us to some kind of amoral relativism? I think not, because no one is *only* a natural or social scientist and no one inhabits solely the third-person perspective. Indeed, the third-person point of view cannot provide a complete understanding of ethical life, since to take it up *by the same token* excludes the perspective of the first person. Certainly we can hope that empirical knowledge will widen and multiply the viewpoints and sympathies available to us. But we should not expect that this will lead us to some position so utterly transcendent that it will secure our ethical intuitions once and for all—granting us a supreme authority that no one else could challenge. If such an external and unassailable position were even possible, I suspect that one would find oneself like a Tibetan monk or Sumbanese aristocrat watching Americans play baseball—curious, perhaps, even knowledgeable, but unlikely to care how the game turns out. But even this would be just part of the story, since the Tibetan and Sumbanese would still have their own games to care about. They too can move between first- and third-person perspectives. Indeed, that capacity to move back and forth between those perspectives (whether or not that capacity is realized in any instance) seems to be a crucial feature of ethical life. As suggested earlier, sometimes people find themselves in the midst of the action, and sometimes they stand apart from it. There is no reason to expect either position to be the final one: the potential for movement between them is an endemic feature of human life.

If empirical knowledge offers no guarantees, it also does not exempt the knowers from being the kind of creatures prone to the taking of ethical stances. To repeat, this is not an optimistic prediction that people are likely to be good. It is an empirical claim that people are oriented to judgments and values, ends in themselves. That orientation may include the perverse or sadistic desire to act against those very values: the ubiquity of ethical life is certainly not the same as the absence of evil. For even those people, acts, and institutions that are called evil involve evaluative dispositions—a sadist would probably be unmotivated and ineffective without them. But participation in ethical life (even for the sadist) is not a choice one could opt out of, short of true pathology. To "opt out" in this sense would not consist in evil but, rather, in utter indifference.

What empirical research might point out is this: neither the first/second person nor the third person on its own fully contains everything we have learned about ethical life. Supposing we were to say that ethics must favor the former and that anyone would rescue their mother or wife, even if she is just the chambermaid, at the expense of the bishop. We begin to have trouble accounting for the two events mentioned in the introduction, the Polish woman who saved the Jewish girl when she could more easily have kept walking by and Sally, who quit her job over the matter of gay marriage when her family depended on her income, and for the British abolitionists in chapter 5 expending time and energy on behalf of people they would never even see—or, for that matter, any young child at all who engages in third-party norm enforcement, as we saw in chapter 1. The self-distancing capacity to step outside of one's first-person position—something that depends on cognitive developments in early childhood—is an affordance frequently taken up in the social history of ethical life. It makes possible the strivings of Urapmin Baptists and the pious Muslims of Cairo, and at least some of the Vietnamese revolutionaries, to radically transform themselves. Yet much of this book demonstrates the limits of the third-person perspective and the morality systems that promote it. The psychological viewpoint shows how high principles are undermined in practice.

Looked at from the perspective of social interaction, they are cross-cut by the constant, inescapable dynamics of respect, recognition, and face-work. The moral torment of the Urapmin is due to their inability to escape the demands of social existence, even when their ideas of Christian morality dub the results immoral. I have suggested that the men who listen to sermons in Cairo often face similar dilemmas. And Don Gabriel found himself caught between loyalty to his son and fealty to the values of his village. If the empirical study of ethical life tells us anything, it should be that these tensions are chronic and not likely to be resolved for good one way or the other. People are endowed with psychological capacities and propensities of which they are mostly unaware and over which they have little or no control. They are embedded in social relationships that are crucial to their sense of self-worth. But they are also purposeful agents who respond to the ideas and arguments their social histories have produced and are prone to contributing new ones. Indeed, it may well be that these very tensions and the impossibility of resolving them once and for all help drive people to make new ethical discoveries and inventions. Without its social histories, ethical life would not be ethical; without its natural histories, it would not be life.

BIBLIOGRAPHY

Agha, Asif. 1997. Tropic aggression in the Clinton–Dole presidential debate. *Pragmatics* 7: 461–498.

———. 2007. *Language and social relations*. Cambridge: Cambridge University Press.

Akin, David, and Joel Robbins. 1999. An introduction to Melanesian currencies: Agency, identity, and social reproduction. In *Money and modernity: State and local currencies in Melanesia*, ed. David Akin and Joel Robbins, 1–40. Pittsburgh, PA: University of Pittsburgh Press.

Anderson, Elizabeth. 1993. *Value in ethics and economics*. Cambridge, MA: Harvard University Press.

———. 2013. Social movements, experiments in living, and moral progress: Case studies from Britain's abolition of slavery. Arthur Allen Leff Fellowship Lecture. Yale Law School, New Haven, CT.

———. Forthcoming. The social epistemology of morality: Learning from the forgotten history of the abolition of slavery. In *The epistemic life of groups*, ed. Miranda Fricker and Michael Brady. Oxford: Oxford University Press.

Anscombe, G.E.M. 1957. *Intention*. Oxford: Basil Blackwell.

———. 1958. Modern moral philosophy. *Philosophy* 33: 1–19.

Antaki, Charles. 1994. *Explaining and arguing: The social organization of accounts*. London: SAGE Publications.

Antze, Paul. 2010. On the pragmatics of empathy in the neurodiversity movement. In *Ordinary ethics: Anthropology, language, and action*, ed. Michael Lambek, 310–327. New York: Fordham University Press.

Appadurai, Arjun. 1996. *Modernity at large: Cultural dimensions of globalization*. Minneapolis: University of Minnesota Press.

Appiah, Anthony. 2008. *Experiments in ethics*. Cambridge, MA: Harvard University Press.

Aristotle. 1941. *Ethica Nichomachea*. Trans. W. D. Ross. In *The basic works of Aristotle*, ed. Richard McKeon, 935–1112. New York: Random House.

Atkinson, J. Maxwell, and John Heritage, eds. 1984. *Structures of social action*. Cambridge: Cambridge University Press.

Austin, J. L. 1979. A plea for excuses. In *Philosophical papers*, ed. J. O. Urmson and G. J. Warnock, 175–204. Oxford: Oxford University Press.

Axelrod, Robert. 1984. *The evolution of cooperation*. New York: Basic Books.

Babb, Lawrence A. 1983. Destiny and responsibility: Karma in popular Hinduism. In *Karma: An anthropological inquiry*, ed. Charles F. Keyes and E. Valentine Daniel, 163–181. Berkeley: University of California Press.

Baird, Jodie A., and Dare A. Baldwin. 2001. Making sense of human behavior: Action parsing and intentional inference. In *Intentions and intentionality*, ed. Bertram F. Malle, Louis J. Moses, and Dare A. Baldwin, 193–206. Cambridge, MA: MIT Press.

Bakhtin, Mikhail. 1981. *The dialogic imagination: Four essays*. Ed. Michael Holquist. Austin: University of Texas Press.

———. 1984. *Problems of Dostoevsky's poetics*. Ed. and trans. Caryl Emerson. Minneapolis: University of Minnesota Press.

Baron-Cohen, Simon. 1995. *Mindblindness: An essay on autism and Theory of Mind*. Cambridge, MA: MIT Press.

Baron-Cohen, Simon, Dare A. Baldwin, and Mary Crowson. 1997. Do children with autism use the speaker's direction of gaze strategy to crack the code of language? *Child Development* 68(1): 48–57.

Bassili, John N. 1989. Trait encoding in behavior identification and dispositional inference. *Personality and Social Psychology Bulletin* 15(3): 285–296.

Bauman, Zygmunt. 1988. *Freedom*. Minneapolis: University of Minnesota Press.

Beidelman, T. O. 1980. The moral imagination of the Kaguru: Some thoughts on tricksters, translation and comparative analysis. *American Ethnologist* 7(1): 27–42.

Benhabib, Seyla. 1992. The generalized and the concrete other: The Kohlberg–Gilligan controversy and moral theory. In *Situating the self: Gender, community and postmodernism in contemporary ethics*, 148–177. New York: Routledge.

Benveniste, Emile. 1971. *Problems in general linguistics*. Trans. Mary Elizabeth Meek. Coral Gables, FL: University of Miami Press.

Besnier, Niko. 1993. Reported speech and affect on Nukulaelae Atoll. In *Responsibility and evidence in oral discourse*, ed. Jane Hill and J. T. Irvine, 161–181. Cambridge: Cambridge University Press.

Bloch, Maurice. 2012. *Anthropology and the cognitive challenge*. Cambridge: Cambridge University Press.

Bloom, Paul. 2004. *Descartes' baby: How the science of child development explains what makes us human*. New York: Basic Books.

Bourdieu, Pierre. 1977. *Outline of a theory of practice*. Trans. Richard Nice. Cambridge: Cambridge University Press.

Bourgois, Philippe. 2003. *In search of respect: Selling crack in El Barrio*. Cambridge: Cambridge University Press.

Bowen, John. 2010. *Why the French don't like headscarves: Islam, the state, and public space*. Princeton, NJ: Princeton University Press.

Bradone, Amanda C., and Henry M. Wellman. 2009. You can't always get what you want: Infants understand failed goal-directed actions. *Psychological Science* 20(1): 85–91.

Brandt, Richard B. 1954. *Hopi ethics: A theoretical analysis*. Chicago: University of Chicago Press.

Bratman, M. E. 1992. Shared cooperative activity. *Philosophical Review* 101(2): 327–341.

Briggs, Jean L. 1970. *Never in anger: Portrait of an Eskimo family*. Cambridge, MA: Harvard University Press.

Brokaw, Cynthia. 1991. *The ledgers of merit and demerit: Social change and moral order in late imperial China*. Princeton, NJ: Princeton University Press.

Brown, Christopher Leslie. 2006. *Moral capital: Foundations of British abolitionism*. Chapel Hill: University of North Carolina Press.

Brown, Roger, and Albert Gilman. 1960. The pronouns of power and solidarity. In *Style in language*, ed. Thomas A. Sebeok, 253–276. Cambridge, MA: MIT Press.

Buechler, Steven M. 1990. *Women's movements in the United States: Woman suffrage, equal rights, and beyond*. New Brunswick, NJ: Rutgers University Press.

Butler, Judith. 2005. *Giving an account of oneself*. New York: Fordham University Press.

Cacioppo, John T., Penny S. Visser, and Cynthia L. Pickett, eds. 2006. *Social neuroscience: People thinking about thinking people*. Cambridge, MA: MIT Press.

Calhoun, Craig. 2012. *The roots of radicalism: Tradition, the public sphere, and early nineteenth-century social movements*. Chicago: University of Chicago Press.

Canguilhem, Georges. 1989. *The normal and the pathological*. Trans. Carolyn R. Fawcett and Robert S. Cohen. New York: Zone Books.

Carr, E. Summerson. 2011. *Scripting addiction: The politics of therapeutic talk and American sobriety*. Princeton, NJ: Princeton University Press.

———. 2013. "Signs of the times": Confession and the semiotic production of inner truth. *Journal of the Royal Anthropological Institute* (N.S.) 19: 34–51.

Caruso, Eugene M., and Francisca Gino. 2011. Blind ethics: Closing one's eyes polarizes moral judgments and discourages dishonest behavior. *Cognition* 118(2): 280–285.

Caton, Steven C. 1986. *Salāam tahīyah*: Greetings from the highlands of Yemen. *American Ethnologist* 13(2): 290–308.

Chandler, Michael J., Bryan W. Sokol, and Darcy Hallett. 2001. Moral responsibility and the interpretive turn: Children's changing conceptions of truth and rightness. In *Intentions and intentionality: Foundations of social cognition*, ed. Bertram F. Malle, Louis J. Moses, and Dare A. Baldwin, 345–365. Cambridge, MA: MIT Press.

Chesebro, James W., J. F. Cragan, and P. McCullough. 1973. The small group technique of the radical revolutionary: A synthetic study of consciousness raising. *Communications Monographs* 40(2): 136–146.

Clark, Eve V. 1997. Conceptual perspective and lexical choice in acquisition. *Cognition* 64: 1–37.

Clark, Herbert H. 2006. Social actions, social commitments. In *Roots of human sociality: Culture, cognition, and human interaction*, ed. N. J. Enfield and Stephen C. Levinson, 126–150. Oxford: Berg Press.

Clifford, James, and George Marcus. 1986. *Writing culture: The poetics and politics of ethnography*. Berkeley: University of California Press.

Cody, Francis. 2013. *The light of knowledge: Literacy activism and the politics of writing in South India*. Ithaca, NY: Cornell University Press.

Cohen, Dov, Brian F. Bowdle, Richard E. Nisbett, and Norbert Schwartz. 1996. Insult, aggression, and the Southern culture of honor: An "experimental ethnography." *Journal of Personality and Social Psychology* 70(5): 945–960.

Csibra, G. 2003. Teleological and referential understanding of action in infancy. *Philosophical Transactions of the Royal Society, London B* 358: 447–358.

Daniel, E. Valentine. 1996. *Charred lullabies: Chapters in an anthropography of violence*. Princeton, NJ: Princeton University Press.

Danziger, Eve. 2006. The thought that counts: Interactional consequences of variation in cultural theories of meaning. In *Roots of human sociality: Culture, cognition, and human interaction*, ed. N. J. Enfield and Stephen C. Levinson, 259–278. Oxford: Berg.

Darwall, Stephen, 1998. *Philosophical ethics*. Boulder, CO: Westview Press.

———. 2006. *The second-person standpoint: Morality, respect, and accountability*. Cambridge, MA: Harvard University Press.

Das, Veena. 2007. *Life and words: Violence and the descent into the ordinary*. Berkeley: University of California Press.

Dave, Naisargi N. 2012. *Queer activism in India: A story in the anthropology of ethics*. Durham, NC: Duke University Press.

Davidson, Donald. 1984. On the very idea of a conceptual scheme. In *Inquiries into truth and interpretation*, 183–198. Oxford: Clarendon Press.

Deacon, Terrence W. 1997. *The symbolic species: The co-evolution of language and the brain*. New York: W. W. Norton.

Descola, Phillippe. 2013. *Beyond nature and culture*. Chicago: University of Chicago Press.

Donham, Donald L. 1999. *Marxist modern: An ethnographic history of the Ethiopian revolution*. Berkeley: University of California Press.

Doniger, Wendy O'Flaherty. 1980. Karma and rebirth in the Vedas and Purāṇas. In *Karma and rebirth in classical Indian traditions*, ed. Wendy O'Flaherty Doniger, 3–37. Berkeley: University of California Press.

Doris, John M. 2002. *Lack of character: Personality and moral behavior*. Cambridge: Cambridge University Press.

Doris, John M., and Alexandra Plakias. 2008. How to argue about disagreement: Evaluative diversity and moral realism. In *Moral psychology, vol. 2: The cognitive science of morality: Intuition and diversity*, ed. Walter Sinnott-Armstrong, 303–331. Cambridge, MA: MIT Press.

Drew, Paul. 1998. Complaints about transgressions and misconduct. *Research on Language and Social Interaction* 31(3–4): 295–325.

Du Bois, John. 2007. The stance triangle. In *Stancetaking in Discourse: Subjectivity, evaluation, interaction*, ed. R. Englebretson, 139–182. Amsterdam: John Benjamins.

Duiker, William J. 1995. *Sacred war: Nationalism and revolution in a divided Vietnam*. New York: McGraw-Hill.

Duranti, Alessandro. 2015. *The anthropology of intention: Language in a world of others*. Cambridge: Cambridge University Press.

Edmonds, David. 2013. *Would you kill the fat man? The trolley problem and what your answer tells us about right and wrong*. Princeton, NJ: Princeton University Press.

Ehrlich, Susan. 2001. *Representing rape: Language and sexual consent*. London: Routledge.

Enfield, N. J. 2013. *Relationship thinking: Agency, enchrony, and human sociality*. Oxford: Oxford University Press.

Englund, Harri. 2008. Extreme poverty and existential obligations: Beyond morality in the anthropology of Africa? *Social Analysis* 52(3): 33–50.

Errington, J. Joseph. 1988. *Structure and style in Javanese: A semiotic view of linguistic etiquette*. Philadelphia: University of Pennsylvania Press.

Evans, E. P. 1906. *The criminal prosecution and capital punishment of animals*. New York: E. P. Dutton.

Evans-Pritchard, E. E. 1937. *Witchcraft, magic and oracles among the Azande*. Oxford: Clarendon Press.

Fader, Ayala. 2009. *Mitzvah girls: Bringing up the next generation of Hasidic Jews in Brooklyn*. Princeton, NJ: Princeton University Press.

Fassin, Didier. 2012. *Humanitarian reason: A moral history of the present.* Berkeley: University of California Press.

Faubion, James. 2001. Toward an anthropology of ethics: Foucault and the pedagogies of autopoiesis. *Representations* 74: 83–104.

———. 2011. *An anthropology of ethics.* Cambridge: Cambridge University Press.

Feuerbach, Ludwig. 1957 (1841). *The essence of Christianity.* Trans. George Eliot. New York: Harper Torchbooks.

Foot, Philippa. 1967. The problem of abortion and the doctrine of the double effect. *Oxford Review* 5: 5–15.

Fortes, Meyer. 1959. *Oedipus and Job in West African religion.* Cambridge: Cambridge University Press.

Foucault, Michel. 1985. *The use of pleasure.* Vol. 2 of *The history of sexuality.* Trans. Robert Hurley. New York: Pantheon.

———. 1997. *Ethics: Subjectivity and truth.* Ed. Paul Rabinow. New York: New Press.

Frantz, Cynthia M., Amy J. C. Cuddy, Molly Burnett, Heidi Ray, and Allen Hart. 2004. A threat in the computer: The race implicit association test as a stereotype threat experience. *Personality and Social Psychology Bulletin* 30(12): 1611–1624.

Freeman, Jo. 1973. The origins of the women's liberation movement. *American Journal of Sociology* 78(4): 792–811.

Freud, Sigmund. 1949. *The ego and the id.* Trans. Joan Riviere. New York: W. W. Norton and Company.

Fuller, C. R. 2004 (1992). *The camphor flame: Popular Hinduism and society in India.* Princeton, NJ: Princeton University Press.

Garfinkel, Harold. 1963. A conception of and experiments with "trust" as a condition of stable concerted actions. In *Motivation and social interaction*, ed. O. J. Harvey, 187–238. New York: Ronald Press.

———. 1967. *Studies in ethnomethodology.* New York: Wiley.

Gershon, Ilana. 2010. Breaking up is hard to do: Media switching and media ideologies. *Journal of Linguistic Anthropology* 20(2): 389–405.

Gibbard, Allan. 1990. *Wise choices, apt feelings: A theory of normative judgment.* Cambridge, MA: Harvard University Press.

Gibson, James J. 1977. The theory of affordances. In *Perceiving, acting, and knowing: Toward an ecological psychology*, ed. R. Shaw and J. Bransford, 67–82. Hillsdale, NJ: Lawrence Erlbaum.

Gilbert, Martin. 2003. *The righteous: The unsung heroes of the Holocaust.* New York: Henry Holt.

Gilligan, Carol. 1982. *In a different voice: Psychological theory and women's development.* Cambridge, MA: Harvard University Press.

Godwin, William. 1793. *Enquiry concerning political justice and its influence on general virtue and happiness, vol. 1.* London: G.G.J. and J. Robinson.

Goffman, Erving. 1959. *The presentation of self in everyday life.* Garden City, NY: Doubleday Anchor.

———. 1963. *Stigma: Notes on the management of spoiled identity.* Englewood Cliffs, NJ: Prentice Hall.

———. 1967. *Interaction ritual: Essays on face-to-face behavior.* Garden City, NY: Anchor Books.

Goodwin, Charles. 2007. Participation, stance and affect in the organization of activities. *Discourse and Society* 18(1): 53–73.

Goodwin, Marjorie. 1990. *He-said-she-said: Talk as social organization among black children*. Bloomington: Indiana University Press.

———. 2006. *The hidden life of girls: Games of stance, status, and exclusion*. Malden, MA: Blackwell Publishers.

Gopnik, Alison, Elizabeth Seiver, and Daphna Buchsbaum. 2013. How causal learning helps us understand other people and how people help us learn about causes: Probabilistic models and the development of social cognition. In *Navigating the social world: What infants, children, and other species can teach us*, ed. Mahzarin R. Banaji and Susan A. Gelman, 186–190. New York: Oxford University Press.

Graham, J., J. Haidt, and B. Nosek. 2009. Liberals and conservatives use different sets of moral foundations. *Journal of Personality and Social Psychology* 96: 1029–1046.

Greene, Joshua, and Jonathan Haidt. 2002. How (and where) does moral judgment work? *Trends in Cognitive Sciences* 6(12): 517–523.

Greene, Joshua D., R. Brian Sommerville, Leigh E. Nystrom, John M. Darley, and Jonathan D. Cohen. 2001. An fMRI investigation of emotional engagement in moral judgment. *Science* 293: 2105–2108.

Greene, Robert H. 2010. *Bodies like bright stars: Saints and relics in Orthodox Russia*. De Kalb: Northern Illinois University Press.

Grice, H. P. 1957. Meaning. *Philosophical Review* 64: 377–388.

———. 1975. Logic and conversation. In *Syntax and semantics, vol. 3: Speech acts*, ed. P. Cole and N. L. Morgan, 41–58. New York: Academic Press.

Gumperz, John J. 1982. *Language and social identity*. New York: Cambridge University Press.

Gupta, Akhil, and James Ferguson, eds. 1997. *Anthropological locations: Boundaries and grounds of a field science*. Berkeley: University of California Press.

Güth, Werner, Rolf Schmittberger, and Bernd Schwarze. 1982. An experimental analysis of ultimatum bargaining. *Journal of Economic Behavior and Organization* 3(4): 367–388.

Hacking, Ian. 1995. *Rewriting the soul: Multiple personality and the sciences of memory*. Princeton, N.J.: Princeton University Press.

———. 2007. *Kinds of people: Moving targets*. Proceedings of the British Academy. Oxford: University of Oxford Press.

Haidt, Jonathan. 2001. The emotional dog and its rational tail: A social intuitionist approach to moral judgment. *Psychological Review* 108(4): 814–834.

———. 2003. The moral emotions. In *Handbook of affective sciences*, ed. Richard J. Davidson, Klaus R. Scherer, and H. Hill Goldsmith, 852–870. Oxford: Oxford University Press.

Hallowell, Irving. 1960. Ojibwa ontology, behavior, and world view. In *Culture in history: Essays in honor of Paul Radin*, ed. Paul Radin and S. Diamond, 19–52. New York: Octagon Books.

Hamlin, J. Kiley, Karen Wynn, and Paul Bloom. 2007. Social evaluation by preverbal infants. *Nature* 450(7169): 557–559.

Hanisch, Carol. 2006. The personal is political (1969): The women's liberation movement classic with a new explanatory introduction. Accessed April 23, 2014. http://www.carolhanisch.org/CHwritings/PersonalisPol.pdf.

———. 2010. Women's liberation consciousness-raising: Then and now. *On the Issues*. Accessed April 23, 2014. http://www.ontheissuesmagazine.com/2010spring/2010spring_Hanisch.php.

Hanks, William. 1992. The indexical ground of deictic reference. In *Rethinking context: Language as an interactive phenomenon*, ed. A. Duranti and C. Goodwin, 43–76. New York: Cambridge University Press.

———. 1996. *Language and communicative practices*. Boulder, CO: Westview Press.

Harkness, Nicholas. 2014. *Songs of Seoul: An ethnography of voice and voicing in Christian South Korea*. Berkeley: University of California Press.

Harris, Sam. 2010. *The moral landscape: How science can determine human values*. New York: Free Press.

———. 2012. Is free will an illusion? Sam Harris on his new book. *Daily Beast*, March 2. Accessed May 30, 2012. http://www.thedailybeast.com/articles/2012/03/02/is-free-will-an-illusion-sam-harris-on-his-new-book.print.html.

Hart, Keith. 2001. Money in an unequal world. *Anthropological Theory* 1(3): 307–330.

Hauser, Marc, Fiery Cushman, Liane Young, R. Kang-Xing Jin, and John Mikhail. 2007. A dissociation between moral judgments and justifications. *Mind and Language* 22(1): 1–21.

Haviland, John. 1977. *Gossip, reputation, and knowledge in Zinacantan*. Chicago: University of Chicago Press.

———. 1997. Shouts, shrieks, and shots: Unruly political conversations in indigenous Chiapas. *Pragmatics* 7: 547–574.

Herodotus. 1997. *Herodotus: The histories*. Trans. George Rawlinson. New York: Alfred A. Knopf.

Herzog, Don. 1998. *Poisoning the minds of the lower orders*. Princeton, NJ: Princeton University Press.

Hill, Jane. 1995. The voices of Don Gabriel: Responsibility and self in a modern Mexicano narrative. In *The dialogic emergence of culture*, ed. Dennis Tedlock and B. Mannheim, 97–147. Urbana: University of Illinois Press.

Hill, Jane, and Judith Irvine. 1992. Introduction. In *Responsibility and evidence in oral discourse*, ed. Jane Hill and Judith Irvine, 1–23. New York: Cambridge University Press.

Hirschfeld, Lawrence A. 2013. The myth of mentalizing and the primacy of folk sociology. In *Navigating the social world: What infants, children, and other species can teach us*, ed. Mahzarin R. Banaji and Susan A. Gelman, 101–106. New York: Oxford University Press.

Hirschfeld, Lawrence A., and Susan A. Gelman. 1997. What young children think about the relationship between language variation and social difference. *Cognitive Development* 12(2): 213–238.

Hirschkind, Charles. 2006. *The ethical soundscape: Cassette sermons and Islamic counterpublics*. New York: Columbia University Press.

Ho, David Yau-fai. 1976. On the concept of face. *American Journal of Sociology* 81(4): 867–884.

Hobson, R. Peter. 1991. Against the theory of "Theory of Mind." *British Journal of Developmental Psychology* 9(1): 33–51.

———. 2004. *The cradle of thought*. New York: Oxford University Press.

Hoffman, M. L. 2000. *Empathy and moral development: Implications for caring and social justice*. New York: Cambridge University Press.

Hollan, Douglas W., and C. Jason Throop. 2011. The anthropology of empathy: Introduction. In *The anthropology of empathy: Experiencing the lives of others in Pacific societies*, ed. Douglas W. Hollan and C. Jason Throop, 1–21. Oxford: Berghahn.

Howell, Signe. 1989. *Society and cosmos: Chewong of peninsular Malaysia*. Chicago: University of Chicago Press.

Hume, David. 1975 (1777). *An enquiry concerning the principles of morals*. Ed. L. A. Selby-Bigge. Oxford: Clarendon Press.

———. 1978 (1740). *A treatise of human nature*. 2nd ed. Ed. L. A. Selby-Bigge. Oxford: Clarendon Press.

Humphrey, Caroline. 1997. Exemplars and rules: Aspects of the discourse of moralities in Mongolia. In *The ethnography of moralities*, ed. Signe Howell, 25–47. London: Routledge.

———. 2012. Inequality. In *A companion to moral anthropology*, ed. Didier Fassin, 302–319. Malden, MA: Wiley-Blackwell.

Hunt, Lynn. 2007. *Inventing human rights: A history*. New York: W. W. Norton.

Husband, William. 2000. *Godless communists: Atheism and society in Soviet Russia, 1917–1932*. De Kalb: Northern Illinois University Press.

Hutchins, Edwin. 1995. *Cognition in the wild*. Cambridge, MA: MIT Press.

Immordino-Yang, Mary Helen, Andrea McColl, Hanna Damasio, and Antonio Damasio. 2009. Neural correlates of admiration and compassion. *Proceedings of the National Academy of Sciences* 106(19): 8021–8026.

Ingold, Tim. 2000. *The perception of the environment: Essays on livelihood, dwelling and skill*. London: Routledge.

Irvine, Judith. 2009. Stance in a colonial encounter: How Mr. Taylor lost his footing. In *Stance: Sociolinguistic perspectives*, ed. A. M. Jaffe, 53–71. New York: Oxford University Press.

Izutsu, Toshihiko. 2002. *Ethico-religious concepts in the Qur'ān*. Montreal: McGill-Queen's University Press.

Jaggar, Alison M. 1989. Love and knowledge: Emotion in feminist epistemology. In *Gender/body/knowledge: Feminist reconstructions of being and knowing*, ed. Alison M. Jaggar and Susan R. Bordo, 145–171. New Brunswick, NJ: Rutgers University Press.

James, Wendy. 1988. *The listening ebony: Moral knowledge, religion, and power among the Uduk of Sudan*. Oxford: Clarendon Press.

Kahneman, Daniel, and Amos Tversky. 1979. Prospect theory: An analysis of decision under risk. *Econometrica* 47(2): 263–291.

Kalish, Charles W. 2006. Integrating normative and psychological knowledge: What should we be thinking about. *Journal of Cognition and Culture* 6(1–2): 191–208.

———. 2012. Generalizing norms and preferences within social categories and individuals. *Developmental Psychology* 48(4): 1133–1143.

Kalish, Charles W., and Christopher A. Lawson. 2008. Development of social category representations: Early appreciation of roles and deontic relations. *Child Development* 79(3): 577–593.

Kant, Immanuel. 1996. Groundwork of the metaphysics of morals (1785). In *Cambridge edition of the works of Immanuel Kant: Practical philosophy*, trans. and ed. Mary J. Gregor, 37–108. Cambridge: Cambridge University Press.

Karmiloff-Smith, Annette. 1992. *Beyond modularity: A developmental perspective on cognitive science*. Cambridge, MA: MIT Press.

Keane, Webb. 1997. *Signs of recognition: Powers and hazards of representation in an Indonesian society*. Berkeley: University of California Press.

———. 2004. Language and religion. In *A companion to linguistic anthropology*, ed. Alessandro Duranti, 431–448. Malden, MA: Blackwell.

———. 2003. Semiotics and the social analysis of material things. *Language and Communication* 23(3): 409–425.

———. 2007. *Christian moderns: Freedom and fetish in the mission encounter*. Berkeley: University of California Press.

———. 2008. Market, materiality, and moral metalanguage. *Anthropological Theory* 8(1): 27–42.

———. 2009. Freedom and blasphemy: On Indonesian press bans and Danish cartoons. *Public Culture* 21(1): 47–76.

———. 2010. Minds, surfaces, and reasons in the anthropology of ethics. In *Ordinary ethics*, ed. Michael Lambek, 64–83. New York: Fordham University Press.

———. 2013. Ontologies, anthropologists, and ethical life. *Hau: Journal of Ethnographic Theory* 3(1): 186–191.

———. 2014. Rotting bodies: The clash of stances toward materiality and its ethical affordances. *Current Anthropology* 55(S10): S312–S321.

Keenan, Elinor Ochs. 1976. The universality of conversational postulates. *Language in Society* 5(1): 67–80.

Kelly, D. J., P. C. Quinn, A. M. Slater, K. Lee, A. Gibson, M. Smith, L. Ge, and O. Pascalis. 2005. Three-month-olds, but not newborns, prefer own-race faces. *Developmental Science* 8(6): F31–F36.

Killen, Melanie, Kelly Lynn Mulvey, Cameron Richardson, Noah Jampol, and Amanda Woodward. 2011. The accidental transgressor: Morally-relevant Theory of Mind. *Cognition* 119(2): 197–215.

Kirsh, W. 1995. The intelligent use of space. *Artificial Intelligence* 73(1): 31–68.

Kitcher, Phillip. 2011. *The ethical project*. Cambridge, MA: Harvard University Press.

Kitzinger, Celia. 2005. Heteronormativity in action: Reproducing the heterosexual nuclear family in after-hours medical calls. *Social Problems* 52(4): 477–498.

Kleinman, Arthur. 1998. From one human nature to many human conditions: An anthropological enquiry into suffering as moral experience in a disordered age. In *Developing anthropological ideas: The Edward Westermarck Lectures 1983–97*, 195–215. Transactions of the Finnish Anthropological Society No. 41. Helsinki.

Kleinman, Arthur, Yunxiang Yan, Jing Jun, Sing Lee, Everett Zhang, Pan Tianshu, Wu Fei, and Guo Jinhua. 2011. *Deep China: The moral life of the person: What anthropology and psychiatry tell us about China today*. Berkeley: University of California Press.

Knappett, Carl. 2005. The affordances of things: A post-Gibsonian perspective on the relationality of mind and matter. In *Rethinking materiality: The engagement of the mind with the material world*, ed. Elizabeth DeMarrais, Chris Gosden, and Colin Renfrew, 43–51. Cambridge: McDonald Institute for Archaeological Research.

Kockelman, Paul. 2004. Stance and subjectivity. *Journal of Linguistic Anthropology* 14(2): 127–150.

Koenigs, Michael, Liane Young, Ralph Adolphs, Daniel Tranel, Fiery Cushman, Marc Hauser, and Antonio Damasio. 2007. Damage to the prefrontal cortex increases utilitarian moral judgements. *Nature* 446(7138): 908–911.

Kohlberg, Lawrence. 1981. *The philosophy of moral development: Moral stages and the idea of justice*. San Francisco: Harper and Row.

Kramer, Bradley H. 2014. Keeping the sacred: Structured silence in the enactment of priesthood authority, gendered worship, and sacramental kinship in Mormonism. Ph.D. diss., University of Michigan, Ann Arbor.

Kremer-Sadlik, Tamar, and Jeemin Kim. 2007. Lessons from sports: Children's socialization to values through family interaction during sports activities. *Discourse and Society* 18(1): 35–52.

Kwon, Heonik. 2006. *After the massacre: Commemoration and consolation in Ha My and My Lai*. Berkeley: University of California Press.

Ladd, John. 1957. *The structure of a moral code: A philosophical analysis of ethical discourse applied to the ethics of the Navaho Indians*. Cambridge, MA: Harvard University Press.

Laidlaw, James. 1995. *Riches and renunciation: Religion, economy, and society among the Jains*. Oxford: Oxford University Press.

———. 2010. Agency and responsibility: Perhaps you can have too much of a good thing. In *Ordinary ethics: Anthropology, language, and action*, ed. Michael Lambek, 143–164. New York: Fordham University Press.

———. 2014. *The subject of virtue: An anthropology of ethics and freedom*. Cambridge: Cambridge University Press.

Lambek, Michael. 2010. Introduction. In *Ordinary ethics: Anthropology, language, and action*, ed. Michael Lambek, 1–36. New York: Fordham University Press.

Lancaster, Roger N. 1994. *Life is hard: Machismo, danger, and the intimacy of power in Nicaragua*. Berkeley: University of California Press.

Larsen, Jeff T., Gary C. Berntson, Kirsten M. Poehlmann, Tiffany A. Ito, and John T. Cacioppo. 2008. The psychophysiology of emotion. In *The handbook of emotions*, 3rd ed., ed. Michael Lewis, Jeannette M. Haviland-Jones, and Lisa Feldman Barrett, 180–195. New York: Guilford.

Lee, Benjamin. 1997. *Talking heads: Language, metalanguage, and the semiotics of subjectivity*. Durham, NC: Duke University Press.

Lempert, Michael. 2012. *Discipline and debate: The language of violence in a Tibetan Buddhist monastery*. Berkeley: University of California Press.

———. 2013. No ordinary ethics. *Anthropological Theory* 13(4): 370–393.

Leshkowich, Ann. 2014. Standardized forms of Vietnamese selfhood: An ethnographic genealogy of documentation. *American Ethnologist* 41(1): 143–162.

Lévi-Strauss, Claude. 1969. *The elementary structures of kinship*. Trans. James Harle Bell and John Richard von Sturmer. Boston: Beacon Press.

Lloyd, G.E.R. 2012. *Being, humanity, and understanding: Studies in ancient and modern societies*. Oxford: Oxford University Press.

Luehrmann, Sonja. 2011. *Secularism Soviet style: Teaching atheism and religion in a Volga republic*. Bloomington: Indiana University Press.

Luhrmann, T. M., R. Padmavati, H. Tharoor, and A. Osei. 2014. Differences in voice-hearing experiences of people with psychosis in the USA, India and Ghana: Interview-based study. *British Journal of Psychiatry*. Accessed July 28, 2014. http://bjp.rcpsych.org/.

Luong, Hy Van. 1988. Discursive practices and power structure: Person-referring forms and sociopolitical struggles in colonial Vietnam. *American Ethnologist* 15(2): 239–253.

———. 1992. *Revolution in the village: Tradition and transformation in North Vietnam, 1925–1988*. Honolulu: University of Hawaii Press.

MacIntyre, Alasdair. 2007. *After virtue: A study in moral theory*. Notre Dame, IN: University of Notre Dame Press.

Madsen, Richard. 1984. *Morality and power in a Chinese village*. Berkeley: University of California Press.

Mahmood, Saba. 2005. *Politics of piety: The Islamic revival and the feminist subject*. Princeton, NJ: Princeton University Press.

———. 2012. Religious freedom, the minority question, and geopolitics in the Middle East. *Comparative Studies in Society and History* 54(2): 418–446.

Malarney, Shaun Kingsley. 2002. *Culture, ritual and revolution in Vietnam*. Honolulu: University of Hawaii Press.

Mandel, Ruth, and Caroline Humphrey. 2002. *Markets and moralities: Ethnographies of postsocialism*. Oxford: Berg.

Manning, Paul. 2008. Barista rants about stupid customers at Starbucks: What imaginary conversations can teach us about real ones. *Language and Communication* 28: 101–126.

Marcus, George E., and Michael M. J. Fischer. 1986. *Anthropology as cultural critique: An experimental moment in the human sciences*. Chicago: University of Chicago Press.

Marr, David. 1971. *Vietnamese anticolonialism, 1885–1925*. Berkeley: University of California Press.

———. 1984. *Vietnamese tradition on trial, 1920–1945*. Berkeley: University of California Press.

Marx, Karl. 1978. The German ideology, part 1 (1846). In *The Marx-Engels reader*, 2nd ed., ed. Robert C. Tucker, 146–200. New York: W. W. Norton.

Mattingly, Cheryl. 2012. Two virtue ethics and the anthropology of morality. *Anthropological Theory* 12(2): 161–184.

Mauss, Marcel. 1990 (1925). *The gift: The form and reason for exchange in archaic societies*. New York: Routledge.

McGovern, Mike. 2013. *Unmasking the state: Making Guinea modern*. Chicago: University of Chicago Press.

McHale, Shawn F. 2004. *Print and power: Confucianism, communism, and Buddhism in the making of modern Vietnam*. Honolulu: University of Hawaii Press.

Mead, George Herbert. 1962 (1934). *Mind, self, and society: From the standpoint of a social behaviorist*. Ed. Charles W. Morris. Chicago: University of Chicago Press.

Meltzoff, Andrew N. 1995. Understanding the intentions of others: Re-enactment of intended acts by 18-month-old children. *Developmental Psychology* 31(5): 838–850.

Meltzoff, Andrew N., and Rechele Brooks. 2001. "Like me" as a building block for understanding other minds: Bodily acts, attention, and intention. In *Intentions and intentionality*, ed. Bertram F. Malle, Louis J. Moses, and Dare A. Baldwin, 171–191. Cambridge, MA: MIT Press.

Meskell, Lynn, and Peter Pels, eds. 2005. *Embedding ethics: Shifting boundaries of the anthropological profession*. London: Berg.

Merritt, Maria. 2000. Virtue ethics and situationist personality psychology. *Ethical Theory and Moral Practice* 3(4): 365–383.

Miller, Daniel. 2001. *The dialectics of shopping*. Chicago: University of Chicago Press.

Miyazaki, Hirokazu. 2013. *Arbitraging Japan: Dreams of capitalism at the end of finance*. Berkeley: University of California Press.

Moll, Henrike, and Michael Tomasello. 2007. Cooperation and human cognition: The Vygotskian intelligence hypothesis. *Philosophical Transactions of the Royal Society B: Biological Sciences* 362(1480): 639–648.

Moran, Richard. 2001. *Authority and estrangement: An essay on self-knowledge*. Princeton, NJ: Princeton University Press.

Mueggler, Erik. 2001. *The age of wild ghosts: Memory, violence, and place in southwest China*. Berkeley: University of California Press.

Nadasdy, Paul. 2007. The gift in the animal: The ontology of hunting and human–animal sociality. *American Ethnologist* 34(1): 25–43.

Neiman, Susan. 2002. *Evil in modern thought: An alternative history of philosophy*. Princeton, NJ: Princeton University Press.

Nietzsche, Friedrich. 1997. *On the genealogy of morality* (1887). Trans. Carol Diethe. Cambridge: Cambridge University Press.

Norman, Donald A. 2002. *The design of everyday things*. New York: Basic Books.

Ochs, Elinor. 1988. *Culture and language development: Language acquisition and language socialization in a Samoan village*. Cambridge: Cambridge University Press.

Ochs, Elinor, and Carolina Izquerdo. 2009. Responsibility in childhood: Three developmental trajectories. *Ethos* 37(4): 391–413.

O'Connor, L. 1969. Defining reality. Mimeograph. Berkeley, CA: Tooth + Nail.

Onishi, Kristine H., and Renée Baillargeon. 2005. Do 15-month-old infants understand false beliefs? *Science* 308(5719): 255–258.

Onvlee, Louis. 1973 (1957). Over de weergave van pneuma. *Cultuur als antwoord*. Leiden: M. Nijhoff.

Oxenfeld, Ellen. 2010. *Drink water, but remember the source: Moral discourse in a Chinese village*. Berkeley: University of California Press.

Pandian, Anand. 2009. *Crooked stalks: Cultivating virtue in South India*. Durham, NC: Duke University Press.

Parfit, Derek. 2011. *On what matters*. Oxford: Oxford University Press.

Parry, Jonathan, and Maurice Bloch, eds. 1989. *Money and the morality of exchange*. Cambridge: Cambridge University Press.

Peris, Daniel. 1998. *Storming the heavens: The Soviet League of the Militant Godless*. Ithaca, NY: Cornell University Press.

Pfaff, Donald W. 2007. *The neuroscience of fair play: Why we (usually) follow the golden rule*. New York: Dana Press.

Piaget, Jean. 1965. *The moral judgment of the child*. Trans. Marjorie Gabain. New York: Free Press.

Pinker, Steven. 2011. *The better angels of our nature: Why violence has declined*. New York: Viking.

Plato. 1961. *Phaedrus*. Trans. R. Hackforth. In *The collected dialogues of Plato, including the letters*, ed. Edith Hamilton and Huntington Cairns, 475–525. Princeton, NJ: Princeton University Press.

Poovey, Mary. 1998. *A history of the modern fact: Problems of knowledge in the sciences of wealth and society*. Chicago: University of Chicago Press.

Porpora, Douglas V., Alexander Nikolaev, Julia Hagemann May, and Alexander Jenkins. 2013. *Post-ethical society: The Iraq war, Abu Ghraib, and the moral failure of the secular*. Chicago: University of Chicago Press.

Povinelli, Elizabeth. 2001. Radical worlds: The anthropology of incommensurability and inconceivability. *Annual Review of Anthropology* 30(1): 319–334.

Putnam, Hilary. 1975. The meaning of "meaning." In *Mind, language and reality. Philosophical papers, vol. 2*, 215–271. Cambridge: Cambridge University Press.

Quine, Willam Van Orman. 1969. *Ontological relativity and other essays*. New York: Columbia University Press.

Rahman, Fazlur. 1994. *Major themes of the Qur'an*. Minneapolis, MN: Bibliotheca Islamica.

Railton, Peter. 1986. Moral realism. *Philosophical Review* 95(2): 163–207.

Rakoczy, Hannes, Felix Warneken, and Michael Tomasello. 2008. The sources of normativity: Young children's awareness of the normative structure of games. *Developmental Psychology* 44(3): 875–881.

Rashdall, Hastings. 1907. *The theory of good and evil, bk. 1: The moral criterion*. Oxford: Oxford University Press.

Rasmussen, Knud. 1929. *Intellectual culture of the Hudson Bay Eskimos*. Copenhagen: Gyldendal.

Rawls, John. 1971. *A theory of justice*. Cambridge, MA: Belknap Press of Harvard University Press.

Read, K. E. 1955. Morality and the concept of the person among the Gahuku-Gama. *Oceania* 25(4): 233–282.

Repacholi, Betty M., and Alison Gopnik. 1997. Early reasoning about desires: Evidence from 14- and 18-month-olds. *Developmental Psychology* 33(1): 12–21.

Rhodes, Marjorie, and Lisa Chalik. 2013. Social categories as markers of intrinsic interpersonal obligations. *Psychological Science* 24(6): 999–1006.

Robbins, Joel. 2004. *Becoming sinners: Christianity and moral torment in a Papua New Guinea society*. Berkeley: University of California Press.

Robbins, Joel, and Alan Rumsey. 2008. Introduction: Cultural and linguistic anthropology and the opacity of other minds. *Anthropological Quarterly* 81(2): 407–420.

Roberts, Elizabeth. 2012. *God's laboratory: Assisted reproduction in the Andes*. Berkeley: University of California Press.

Rogers, Douglas. 2009. *The Old Faith and the Russian land: A historical ethnography of ethics in the Urals*. Ithaca, NY: Cornell University Press.

Rosaldo, Michelle Z. 1982. The things we do with words: Ilongot speech acts and speech act theory in philosophy. *Language in Society* 11(2): 203–237.

Rosaldo, Renato. 1989. *Culture and truth: The remaking of social analysis*. Boston: Beacon Press.

Rosen, Michael. 2012. *Dignity: Its history and meaning*. Cambridge, MA: Harvard University Press.

Rosenthal, Naomi Braun. 1984. Consciousness raising: From revolution to re-evaluation. *Psychology of Women Quarterly* 8(4): 309–326.

Ross, L. 1977. The intuitive psychologist and his shortcomings: Distortions in the attribution process. In *Advances in experimental social psychology, vol. 10*, ed. L. Berkowitz, 173–220. New York: Academic Press.

Ross, Lee, and Richard E. Nisbett. 1991. *The person and the situation: Perspectives of social psychology*. New York: McGraw-Hill Book Company.

Rozin, Paul, and Carol Nemeroff. 1990. The laws of sympathetic magic. In *Cultural psychology: Essays on comparative human development*, ed. James W. Stigler, Richard A. Shweder, and Gilbert Herdt, 205–232. New York: Cambridge University Press.

Rozin, Paul, and Edward B. Royzman. 2001. Negativity bias, negativity dominance, and contagion. *Personality and Social Psychology Review* 5(4): 296–320.

Rumsey, Alan. 2003. Language, desire, and the ontogenesis of intersubjectivity. *Language and Communication* 23(2): 169–187.

Sacks, Harvey. 1972. Notes on police assessment of moral character. In *Studies in social interaction*, ed. David N. Sudnow, 280–293. New York: Free Press.

———. 1984. On doing "being ordinary." In *Structures of social action: Studies in conversation analysis*, ed. J. Maxell Atkinson and John Heritage, 413–429. Cambridge: Cambridge University Press.

Sacks, Harvey, Emanuel A. Schegloff, and Gail Jefferson. 1974. A simplest systematics for the organization of turn-taking for conversation. *Language* 50: 696–735.

Sahlins, Marshall. 1985. *Islands of history*. Chicago: University of Chicago Press.

Sandel, Michael J. 2012. *What money can't buy: The moral limits of markets*. New York: Farrar, Straus and Giroux.

Sarachild, Kathie. 1978. Consciousness-raising: A radical weapon. *Feminist Revolution* 144: 145–146. Accessed April 20, 2014. http://library.duke.edu/rubenstein/scriptorium/wlm.fem/sarachild.html.

Schachter, Stanley, and Jerome Singer. 1962. Cognitive, social and psychological determinants of emotional state. *Psychological Review* 69: 379–399.

Schegloff, Emanuel. 2006. Interaction: The infrastructure for social institutions, the natural ecological niche for language, and the arena in which culture is enacted. In *Roots of human society: Culture, cognition and interaction*, ed. N. J. Enfield and Stephen C. Levinson, 70–96. Oxford: Berg.

Scheman, Naomi. 1980. Anger and the politics of naming. In *Women and language in literature and society*, ed. Sally McConnell-Ginet, Ruth Borker, and Nelly Furman, 174–188. New York: Praeger.

Schieffelin, Bambi B. 1990. *The give and take of everyday life: Language socialization of Kaluli children*. Cambridge: Cambridge University Press.

———. 2008. Speaking only your own mind: Reflections on talk, gossip and intentionality in Bosavi (PNG). *Anthropological Quarterly* 81(2): 431–441.

Schielke, Samuli. 2009. Being good in Ramadan: Ambivalence, fragmentation, and the moral self in the lives of young Egyptians. *Journal of the Royal Anthropological Institute* 15(S1): S24–S40.

Schneewind, Jerome B. 1998. *The invention of autonomy: A history of modern moral philosophy*. New York: Cambridge University Press.

Schutz, Alfred. 1967 (1932). *The phenomenology of the social world*. Trans. George Walsh and Frederick Lehnert. Evanston, IL: Northwestern University Press.

Schwartz, Norbert, and Gerd Bohner. 2001. The construction of attitudes. In *Blackwell handbook of social psychology: Intraindividual processes*, ed. Abraham Tesser and Norbert Schwarz, 436–457. Oxford: Blackwell Publishers.

Scott, James C. 1976. *The moral economy of the peasant: Rebellion and subsistence in Southeast Asia*. New Haven, CT: Yale University Press.

Scott, Joan W. 1991. The evidence of experience. *Critical Inquiry* 17(4): 773–797.

Shiraishi, Takashi. 1990. *An age in motion: Popular radicalism in Java, 1912–1926*. Ithaca, NY: Cornell University Press.

Shryock, Andrew. 2012. Breaking hospitality apart: Bad hosts, bad guests, and the problem of sovereignty. *Journal of the Royal Anthropological Institute* 18(S1): S20–S33.

Shweder, Richard A. 1990. Ethical relativism: Is there a defensible version? *Ethos* 18(2): 205–218.

Shweder, Richard A., Manamohan Mahapatra, and Joan G. Miller. 1990. Culture and moral development. In *Cultural psychology: Essays on comparative human development*, ed. James W. Stigler, Richard A. Shweder, and Gilbert Herdt, 130–204. New York: Cambridge University Press.

Sidnell, Jack. 2010a. *Conversation analysis: An introduction*. Malden, MA: Wiley-Blackwell.

———. 2010b. The ordinary ethics of everyday talk. In *Ordinary ethics: Anthropology, language, and action*, ed. Michael Lambek, 123–139. New York: Fordham University Press.

———. 2014. Interaction and intersubjectivity. In *The Cambridge handbook of linguistic anthropology*, ed. N. J. Enfield, Paul Kockelman, and Jack Sidnell, 364–399. Cambridge: Cambridge University Press.

Silverstein, Michael. 1993. Metapragmatic discourse and metapragmatic function. In *Reflexive language: Reported speech and metapragmatics*, ed. John Lucy, 33–58. Cambridge: Cambridge University Press.

———. 2003a. Indexical order and the dialectics of sociolinguistic life. *Language and Communication* 23(3): 193–229.

———. 2003b. Translation, transduction, transformation: Skating "glossando" on thin semiotic ice. In *Translating cultures: Perspectives on translation and anthropology*, ed. Paula G. Rubel and Abraham Rosman, 75–105. Oxford: Berg.

Silverstein, Michael, and Greg Urban, eds. 1996. *Natural histories of discourse*. Chicago: University of Chicago Press.

Singer, Peter. 1981. *The expanding circle: Ethics and sociobiology*. New York: Farrar Strauss Giroux.

Slater, Alan, Charlotte Von der Schulenburg, Elizabeth Brown, Marion Badenoch, George Butterworth, Sonia Parsons, and Curtis Samuels. 1998. Newborn infants prefer attractive faces. *Infant Behavior and Development* 21(2): 345–354.

Slingerland, Edward. 2007. *Effortless action:* Wu-wei *as conceptual metaphor and spiritual ideal in early China*. Oxford: Oxford University Press.

Smith, Adam. 1976 (1790). *The theory of moral sentiments*. Ed. D. D. Rafael and A. L. Macfie. Oxford: Oxford University Press.

Smith, Wilfred Cantwell. 1962. *The meaning and end of religion: A new approach to the religious traditions of mankind*. New York: Macmillan.

Sondheim, Stephen. 1957. Gee, Officer Krupke. *West Side Story*. Leonard Bernstein Music Publishing Company LLC.

Sperber, Dan. 2000. Metarepresentation in an evolutionary perspective. In *Metarepresentations: A multidisciplinary approach*, ed. Dan Sperber, 117–138. Oxford: Oxford University Press.

Stafford, Charles, ed. 2013. *Ordinary ethics in China*. LSE Monographs on Social Anthropology. London: Bloomsbury Academic.

Stasch, Rupert. 2008. Knowing minds is a matter of authority: Political dimensions of opacity statements in Korowai moral psychology. *Anthropological Quarterly* 81(2): 443–454.

Steele, Claude M., and Joshua Aronson. 1995. Stereotype threat and the intellectual test performance of African Americans. *Journal of Personality and Social Psychology* 69(5): 797–811.

Steinem, Gloria. 1995. *Outrageous acts and everyday rebellions*. New York: H. Holt.

Stone, Valerie E. 2006. Theory of Mind and the evolution of social intelligence. In *Social neuroscience: People thinking about thinking people*, ed. John T. Cacioppo, Penny S. Visser, and Cynthia L. Pickett, 103–129. Cambridge, MA: MIT Press.

Strawson, P. F. 1974. *Freedom and resentment: And other essays*. London: Methuen.

Sullivan, Winnifred Fallers. 2005. *The impossibility of religious freedom*. Princeton, NJ: Princeton University Press.

Sweetser, Eve E. 1987. The definition of lie. In *Cultural models in language and thought*, ed. Dorothy Holland and Naomi Quinn, 43–66. Cambridge: Cambridge University Press.

Tai, Hue-Tam Ho. 1983. *Millenarianism and peasant politics in Vietnam*. Cambridge, MA: Harvard University Press.

Tannen, Deborah. 2005. *Conversational style: Analyzing talk among friends*. Oxford: Oxford University Press.

Taylor, Charles. 1985. *Human agency and language*. Cambridge: Cambridge University Press.

Taylor, K. W. 2013. *A history of the Vietnamese*. Cambridge: Cambridge University Press.

Taylor, Philip. 2004. *Goddess on the rise: Pilgrimage and popular religion in Vietnam*. Honolulu: University of Hawaii Press.

Theidon, Kimberly. 2013. *Intimate enemies: Violence and reconciliation in Peru*. Philadelphia: University of Pennsylvania Press.

Thompson, E. P. 1971. The moral economy of the English crowd in the eighteenth century. *Past and Present* 50: 76–136.

Thomson, Judith Jarvis. 1976. Killing, letting die, and the trolley problem. *Monist* 59(2): 204–217.

Tomasello, Michael. 1999. *The cultural origins of human cognition*. Cambridge, MA: Harvard University Press.

———. 2009. *Why we cooperate*. Cambridge, MA: MIT Press.

Tomasello, Michael, Malinda Carpenter, Josep Call, Tanya Behne, and Henrike Moll. 2005. Understanding and sharing intentions: The origins of cultural cognition. *Behavioral and Brain Science* 28: 675–691.

Trevarthen, Colwyn, and Kenneth J. Aitken. 2001. Infant intersubjectivity: Research, theory, and clinical applications. *Journal of Child Psychology and Psychiatry* 42(1): 3–48.

Trivers, Robert L. 1971. The evolution of reciprocal altruism. *Quarterly Review of Biology* 46(1): 35–57.

———. 1991. Deceit and self-deception: The relationship between communication and consciousness. In *Man and beast revisited*, ed. M. Robinson and L. Tiger, 175–191. Washington, DC: Smithsonian.

Turiel, Elliot. 1983. *The development of social knowledge: Morality and convention*. Cambridge: Cambridge University Press.

Valeri, Valerio. 2000. *The forest of taboos: Morality, hunting, and identity among the Huaulu of the Moluccas*. Madison: University of Wisconsin Press.

Velleman, David. 2013. *Foundations for moral relativism*. Cambridge: Open Book Publishers.

Verplaetse, Jan, Jelle De Schrijver, Sven Vanneste, and Johan Braeckman, eds. 2009. *The moral brain: Essays on the evolutionary and neuroscientific aspects of morality*. New York: Springer.

Vivieros de Castro, Eduardo. 1998. Cosmological deixis and Amerindian perspectivism. *Journal of the Royal Anthropological Institute* 4(3): 469–488.

Voloshinov, V. N. 1986. *Marxism and the philosophy of language*. Cambridge, MA: Harvard University Press.

Vygotsky, L. S. 1978. *Mind and society: The development of higher mental processes*. Ed. Michael Cole, Vera John-Steiner, Sylvia Scribner, and Ellen Souberman. Cambridge, MA: Harvard University Press.

Wallach, Wendell, Stan Franklin, and Colin Allen. 2009. A conceptual and computational model of moral decision making in human and artificial agents. *Topics in Cognitive Science* 2: 454–485.

Warneken, Felix, and Michael Tomasello. 2006. Altruistic helping in human infants and young chimpanzees. *Science* 311(5765): 1301–1303.

———. 2009. Varieties of altruism in children and chimpanzees. *Trends in Cognitive Sciences* 12: 397–482.

Warner, Michael. 1999. *The trouble with normal: Sex, politics, and the ethics of queer life*. New York: Free Press.

Weber, Max. 1978 (1920). *The Protestant ethic and the spirit of capitalism*. Trans. Talcott Parsons. New York: Charles Scribner.

Welker, Marina. 2014. *Enacting the corporation: An American mining firm in post-authoritarian Indonesia*. Berkeley: University of California Press.

Wellman, Henry M. 1992. *The child's Theory of Mind*. Cambridge, MA: MIT Press.

Wellman, Henry M., and David Liu. 2004. Scaling of Theory-of-Mind tasks. *Child Development* 75(2): 523–541.

Wertsch, James V. 1991. *Voices of the mind: Sociocultural approach to mediated action*. Cambridge, MA: Harvard University Press.

———. 1998. Mediated action. In *A companion to cognitive science*, ed. William Bechtel and George Graham, 518–525. Malden, MA: Blackwell.

Widdicombe, Sue. 1993. Autobiography and change: Rhetoric and authenticity of "Gothic" style. In *Discourse analytic research: Repertoires and readings of texts in action*, ed. E. Burman and I. Parker, 94–113. London: Routledge.

Williams, Bernard. 1981. *Moral luck: Philosophical papers 1973–1980*. Cambridge: Cambridge University Press.

———. 1985. *Ethics and the limits of philosophy*. Cambridge, MA: Harvard University Press.

Wilson, Edward O. 1976. The social instinct. *Bulletin of the American Academy of Arts and Sciences* 30(1), October: 11–25.

Wilson, Marisa. 2014. *Everyday moral economies: Food, politics and scale in Cuba*. Malden, MA: Wiley-Blackwell.

Wittgenstein, Ludwig. 1953. *Philosophical investigations*. Trans. G.E.M. Anscombe. Oxford: B. Blackwell.

Woodward, Amanda L. 1998. Infants selectively encode the goal object of an actor's reach. *Cognition* 69(1): 1–34.

Zelizer, Viviana A. 2005. *The purchase of intimacy*. Princeton, NJ: Princeton University Press.

Zigon, Jarrett. 2007. Moral breakdown and the ethical demand: A theoretical framework for an anthropology of moralities. *Anthropological Theory* 7(2): 131–150.

Zimmerman, Don H. 1992. The interactional organisation of calls for emergency assistance. In *Talk at work: Interaction in institutional settings*, ed. Paul Drew and John Heritage, 418–469. Cambridge: Cambridge University Press.

Zunshine, Lisa. 2006. *Why we read fiction: Theory of Mind and the novel*. Columbus: Ohio University Press.

Index

Lightning Source UK Ltd.
Milton Keynes UK
UKHW012202230120
357502UK00002B/244